AF352819

Atmospheric Composition and Cloud Cover Observations

Atmospheric Composition and Cloud Cover Observations

Editor

Mirela Voiculescu

MDPI • Basel • Beijing • Wuhan • Barcelona • Belgrade • Manchester • Tokyo • Cluj • Tianjin

Editor
Mirela Voiculescu
Universitatea Dunarea de Jos Galati
Romania

Editorial Office
MDPI
St. Alban-Anlage 66
4052 Basel, Switzerland

This is a reprint of articles from the Special Issue published online in the open access journal *Atmosphere* (ISSN 2073-4433) (available at: https://www.mdpi.com/journal/atmosphere/special_issues/atmo_composition).

For citation purposes, cite each article independently as indicated on the article page online and as indicated below:

LastName, A.A.; LastName, B.B.; LastName, C.C. Article Title. *Journal Name* **Year**, *Volume Number*, Page Range.

ISBN 978-3-0365-0824-5 (Hbk)
ISBN 978-3-0365-0825-2 (PDF)

Contents

About the Editor

Mirela Voiculescu has a doctoral degree in physics and her present position is Professor at "Dunarea de Jos" University of Galati, Romania. Her research focuses on ionospheric physics, atmospheric and meteorological processes, near-Earth environments, and solar–terrestrial interactions, with a focus on the challenging subject of disentangling sources of climate variability. She teaches courses on atmospheric physics, meteorology, climate variability, and remote sensing at undergraduate and graduate level.

Editorial

Special Issue Atmospheric Composition and Cloud Cover Observations

Mirela Voiculescu

Faculty of Sciences and Environment, University Dunarea de Jos, ECEE, Str. Domneasca nr 111, 800201 Galati, Romania; Mirela.Voiculescu@ugal.ro

Received: 23 December 2020; Accepted: 30 December 2020; Published: 31 December 2020

A Special Issue of *Atmosphere*, "*Atmospheric Composition and Cloud Cover Observations*", is focused on presenting some of the latest results of observations of clouds and atmospheric composition, mainly by referring to new equipment or experimental set-ups. The seven articles describe observational results using remote sensing and in situ methods that observe cloud cover and/or atmospheric composition, and means of using modelling for evaluating the state of the *Atmosphere* at local, regional, or global scale for various time periods. Most articles focused on observing the atmospheric composition (four), one describes a possibility for obtaining information about cloud cover as a by-product of cameras, one compares model and observational results for some meteorological parameters, and the remaining one analyzes a relatively rare but damaging meteorological hazard.

The diurnal and day-to-day variability of some atmospheric constituents (NO_x, SO_2, CO, O_3, CH_4) during winter, around the capital of Romania, Bucharest, was investigated in Marin et al. [1] using in-situ observations. Differences between work days and weekends were identified. They have also linked the concentration of some air pollutants with some hot spots, using coincident meteorological (mainly wind) data. They conclude that SO_2 may be transported at long distances, while CO and NO_x mainly originate in relatively close traffic hot spots and residential heating. Another important finding is that, locally, air quality is good for the European standards.

The aerosol composition over Europe between 2008 and 2018 and the dependence on atmospheric circulation types were analyzed by Nicolae et al. [2] using EARLINET and AERONET databases. They showed that smoke and dust particles are predominant in Europe in most years and that there is no clear seasonal dependence, but there is a significant variation with the year which can most likely be associated with various circulation patterns. They also analyzed the spatial distribution over the continent in an attempt to identify whether there is any trend in the aerosol composition and/or variability.

Results of the Weather Research and Forecasting (WRF) model were discussed by Roșu et al. [3] in connection with observational data collected by a lidar platform, photometer and meteorological instruments at a local site. These helped in evaluating the performance of simulating the Planetary Boundary Layer Height (PBLH), temperature, wind speed, pressure at the surface and total precipitable water vapor. They found that the model performed relatively well in most cases and that differences from measured parameters are not uncommon.

Effects of biomass burning (mostly wildfires) on the atmospheric composition were described by Guo et al. [4], using remote sensing data from Aqua MODIS (MODerate-resolution Imaging Spectroradiometer). These were combined with CO_2 results from the OCO-2 model, which uses results of the Orbiting Carbon Observatory-2, which helped in identifying advantages and disadvantages of the model. They conclude that OCO-2 is a useful tool for fire-emission monitoring and needs to be used at a larger scale for better management of wildfire effects on atmospheric composition.

Occurrence and climatology of an unpleasant meteorological hazard that may cause important perturbations and damages, the freezing rain, was examined by Andrei et al. [5]. They used reanalysis ERA5 data and radiosonde observations for analyzing a particular case in the SE of Europe, but also for building a climatology of these events. They concluded that specific synoptic circulation patterns, involving Scandinavian or Siberian anticyclones, together with a yet unclear role of the Carpathians, significantly increase the probability that rain freezing events occur in Romania.

Observational characteristics of cloud cover at a local site in France are described by Baray et al. [6], who describe an algorithm developed for calculating cloud occurrence frequency using images of an automatic camera. The climatology of frequency occurrence is compared with larger scale products derived from reanalysis and remote sensing, with good results in terms of diurnal and seasonal variability. Absolute values of cloud frequency are less accurate; however, the authors make a good case for using camera images in areas with no cloud instrumentation.

The most recent article provides a multi-instrumental analysis of a dust outbreak over a European region. Ajtai et al. [7] combine remote data from a sun photometer and lidar with in situ measurements of a particle counter and HYSPLIT trajectory analysis, in order to identify the Saharan origin of the dust event observed locally at a site in the SE of Europe. The article concludes that continuous monitoring over larger areas helps in forecasting and managing such hazardous events, which are not rare and occur in all seasons.

Funding: This research received no external funding.

Acknowledgments: Thanks are expressed to the authors for their contributions.

Conflicts of Interest: The author of this editorial declares no conflict of interest.

References

1. Marin, C.A.; Mărmureanu, L.; Radu, C.; Dandocsi, A.; Stan, C.; Ţoancă, F.; Preda, L.; Antonescu, B. Wintertime Variations of Gaseous Atmospheric Constituents in Bucharest Peri-Urban Area. *Atmosphere* **2019**, *10*, 478. [CrossRef]

2. Nicolae, V.; Talianu, C.; Andrei, S.; Antonescu, B.; Ene, D.; Nicolae, D.; Dandocsi, A.; Toader, V.-E.; Stefan, S.; Savu, T.; et al. Multiyear Typology of Long-Range Transported Aerosols over Europe. *Atmosphere* **2019**, *10*, 482. [CrossRef]

3. Rosu, I.-A.; Ferrarese, S.; Radinschi, I.; Ciocan, V.; Cazacu, M.-M. Evaluation of Different WRF Parametrizations over the Region of Iași with Remote Sensing Techniques. *Atmosphere* **2019**, *10*, 559. [CrossRef]

4. Guo, M.; Li, J.; Wen, L.; Huang, S. Estimation of CO_2 Emissions from Wildfires Using OCO-2 Data. *Atmosphere* **2019**, *10*, 581. [CrossRef]

5. Andrei, S.; Antonescu, B.; Boldeanu, M.; Mărmureanu, L.; Marin, C.A.; Vasilescu, J.; Ene, D. An Exceptional Case of Freezing Rain in Bucharest (Romania). *Atmosphere* **2019**, *10*, 673. [CrossRef]

6. Baray, J.-L.; Bah, A.; Cacault, P.; Sellegri, K.; Pichon, J.-M.; Deguillaume, L.; Montoux, N.; Noel, V.; Sèze, G.; Gabarrot, F.; et al. Cloud Occurrence Frequency at Puy de Dôme (France) Deduced from an Automatic Camera Image Analysis: Method, Validation, and Comparisons with Larger Scale Parameters. *Atmosphere* **2019**, *10*, 808. [CrossRef]

7. Ajtai, N.; Stefanie, H.; Mereuţă, A.; Radovici, A.; Botezan, C.S. Multi-Sensor Observation of a Saharan Dust Outbreak over Transylvania, Romania in April 2019. *Atmosphere* **2020**, *11*, 364. [CrossRef]

Article

Multi-Sensor Observation of a Saharan Dust Outbreak over Transylvania, Romania in April 2019

Nicolae Ajtai, Horațiu Ștefănie, Alexandru Mereuță, Andrei Radovici and Camelia Botezan *

Faculty of Environmental Science and Engineering, Babeș-Bolyai University, 30 Fântânele St.,
400294 Cluj-Napoca, Romania; nicolae.ajtai@ubbcluj.ro (N.A.); horatiu.stefanie@ubbcluj.ro (H.Ș.);
mereutaalexandru90@gmail.com (A.M.); radovici_andrei@yahoo.com (A.R.)
* Correspondence: camelia.costan@ubbcluj.ro

Received: 17 March 2020; Accepted: 8 April 2020; Published: 9 April 2020

Abstract: Mineral aerosols are considered to be the second largest source of natural aerosol, the Saharan desert being the main source of dust at global scale. Under certain meteorological conditions, Saharan dust can be transported over large parts of Europe, including Romania. The aim of this paper is to provide a complex analysis of a Saharan dust outbreak over the Transylvania region of Romania, based on the synergy of multiple ground-based and satellite sensors in order to detect the dust intrusion with a higher degree of certainty. The measurements were performed during the peak of the outbreak on April the 24th 2019, with instruments such as a Cimel sun-photometer and a multi-wavelength Raman depolarization lidar, together with an in-situ particle counter measuring at ground level. Remote sensing data from MODIS sensors on Terra and Aqua were also analyzed. Results show the presence of dust aerosol layers identified by the multi-wavelength Raman and depolarization lidar at altitudes of 2500–4000 m, and 7000 m, respectively. The measured optical and microphysical properties, together with the HYSPLIT back-trajectories, NMMB/BSC dust model, and synoptic analysis, confirm the presence of lofted Saharan dust layers over Cluj-Napoca, Romania. The NMMB/BSC dust model predicted dust load values between 1 and 1.5 g/m^2 over Cluj-Napoca at 12:00 UTC for April the 24th 2019. Collocated in-situ PM monitoring showed that dry deposition was low, with PM10 and PM2.5 concentrations similar to the seasonal averages for Cluj-Napoca.

Keywords: Saharan dust; aerosol remote sensing; lidar; sun-photometer; MODIS

---

1. Introduction

Aerosols are an important component of the atmospheric mixture which influence radiative forcing through the atmosphere [1]. Besides the radiative effects, aerosols can pose a significant hazard to human health [2,3] and to certain economic sectors, having a significant economic impact. In this case, similar to what took place in some Eastern and Western Asian countries, aerosols can affect constructions, aerial and terrestrial transportation (flight delay and traffic accident rate changes), trade, agriculture (crop degradation), manufacturing, and households over short and long term periods [4,5].

Mineral aerosols are considered to be the second largest source of natural aerosol [6]. They usually originate from soil ablation in arid areas and are lofted to high altitudes by thermal turbulences and are subjected to long range transport [7,8]. Dust originating from the Sahara and Sahel regions of Northern Africa accounts for more than 50% of the global atmospheric mineral dust [9,10]. It is estimated that roughly 800 million tones/year of dust is subjected to aeolian transport from these two sources alone [11,12]. Upwards of 10% of these Saharan dust emissions are transported over continental Europe and the Mediterranean Sea [11]. Seasonal trajectories have been discussed in the scientific literature, with peak emissions observed in spring and summer [12,13]. Studies such as [14] show that transport over larger ranges are expected during April through June, while vertical transport is highest from April through September. Under certain meteorological conditions, significant amounts of Saharan dust are transported north, over the Mediterranean basin [15–17] and in severe cases, even over Western, Central, and Eastern Europe [18–21]. Dust outbreaks have been observed as far as northern Europe and the Baltic Sea [10,22]. Dust trajectories from North Africa to Continental Europe are often linked to the movement of two pressure systems. The strength and dynamics of these systems determine the potential to carry dust plumes over long ranges [22,23].

A severe dust intrusion can have intense socio-economic and environmental consequences. In previous studies, it was observed that the amount of Saharan dust in Romania can be above the threshold presented in the national legislation, which poses danger for human health [24]. Furthermore, a Saharan dust plume can influence the climate of Romania. Dust particles lofted above Romania are majorly mixed with pollutants of other European countries, and can easily lead to air pollution [25]. Concerning the additional environmental impacts of a dust intrusion, the pH of precipitations can increase due to the influence of Saharan dust events, similar to what took place in other European countries [26].

There is a need, at stakeholder level, for near-real-time monitoring and early warning systems. These systems could be developed for a region as extensive as Romania with the input of additional data sets from other sources [27]. With the integration of different data sets, information on the size and concentration of the dust particles, the dynamics of the dust plume, its altitude, etc. could be obtained. Generally, using different data from various sensors gives us the ability to comprehensively identify the dust intrusion's characteristics (that could not be performed using just a single sensor). The more information is sent to stakeholders, policy makers, and natural hazards specialists, the more accurate their decisions will become.

Therefore, in this paper, our main intention is to provide more information through multi-source data integration. In Romania there are several studies focusing on Saharan dust intrusions [28–31]. However, these studies were conducted in the southern and eastern part of Romania. In the north-western part of Romania, there is limited information on Saharan dust intrusions, with the majority of the studies carried out in this region [24,32,33] relying mostly on single-sensor information.

Due to the high degree of heterogeneity and spatial-temporal variability of atmospheric aerosols, the study of their composition, distribution, and dynamics is currently monitored at a global level. This monitoring can be ground-based, airborne or satellite-based.

Talianu et al., 2007 [34] detected a dust event over Bucharest, Romania, and also determined the dust type using a lidar system accompanied by a model data prognosis and air masses backward trajectories. Tudose, 2013 [25] studied a Saharan dust intrusion in Romania (occurred in July 2012) and its influences on the local weather using lidar data and complementary methods. Ajtai et al., 2017, [32] found that the aerosol properties over Cluj-Napoca, Romania derived from AERONET sun-photometric data using a CIMEL CE 318A radiometer. Labzovskii et al., 2014 [29] attempted to determine the properties of a Saharan dust intrusion (occurred in May and June 2013) over Bucharest (south of Romania) using lidar and sun-photometer. Ştefănie et al., 2015 [24] detected the dust intrusion over Cluj-Napoca, Romania (in April 2014) using a lidar system, FLEXPART and HYSPLIT models. They identified the source of the dust plume and measured the ground level concentration of dust particles using an optical aerosol monitor. Cazacu et al., 2017 [35] assessed the temporal and vertical variation of Saharan dust concentration near Bucharest, Romania (from the data related to 2012) using the lidar system and some auxiliary methods. Mărmureanu et al., 2019 [36] studied the synoptic-scale conditions leading to the Saharan dust intrusion that was mixed with snow precipitation (occurred in March 2018) over the south-east of Romania. They analysed the dust morphology, particle size, and the physical and chemical properties. They also developed a method that visualized the presence of some minerals.

This paper contributes to the above-mentioned regional studies analyzing Saharan dust intrusions in the north-western part of Romania (which have not yet been sufficiently studied in this part of the country) using integrated, prevalent techniques and equipment types and satellite-based imagery. Thus, the integration of different datasets brings added value and leads to the extraction of valuable and more reliable information on the phenomenon of dust intrusions.

The aim of this paper is to provide a complex analysis of a Saharan dust outbreak over the Transylvania region of Romania based on the synergy of multiple ground-based and satellite sensors, in order to detect the dust intrusion with a higher degree of certainty. This outbreak occurred between April 21st and 25th 2019, affecting large areas of the European continent, reaching as far north as Iceland and Norway [37]. The exceptionally dust event was generated by a "haboob"- a cold low-pressure system, developed on 20th of April 2019 over north-west Africa (Morocco and Algeria). Significant amounts of dust were lifted up into the upper troposphere and then advected by the persistent southern circulation, at first towards Western Europe and western Mediterranean, and later on (starting on 23rd of April 2019) towards south-eastern Europe. Dust loaded air masses arrived over Transylvania, Romania on April the 24th 2019.

2. Instruments and Method

We analyzed data measured in Cluj-Napoca, Romania on April the 24th 2019 with a series of ground-based remote sensing instruments, such as multi-wavelength Raman depolarization lidar, part of the European Aerosol Research Lidar Network (EARLINET) [38], and a sun-photometer, part of the Aerosol Robotic Network (AERONET) [39]. Currently, the two networks are part of the Aerosols, Clouds, and Trace gases Research InfraStructure (ACTRIS) [40]. The Cluj-Napoca ACTRIS station can be considered representative for the whole Transylvanian territory, and also for the Pannonian Basin, given the limited coverage of AERONET and EARLINET in Hungary, Austria, and south-western Ukraine. Besides the ground-based instruments, remote sensing data from MODIS sensors on Terra and Aqua satellites were used in this analysis. During the intrusion, an in-situ optical particle counter measured particulate matter concentrations at ground level to account for potential dust deposition.

Each data source has some limitations. For example, the altitude information is not included in the MODIS product, but can be inferred from the lidar data. On the other hand, the two-dimensional distribution of dust is inferred by the MODIS images. By integrating these single sensor data products, new added value information will be extracted which could not have been collected by a single sensor.

2.1. Cimel CE 318 Sun Photometer

The Cimel CE 318 sun-photometer is a ground-based sun and sky tracking automated radiometer, which measures aerosol optical properties using a combination of filters and azimuth and zenith mobility. It measures sun and sky radiance in order to derive the aerosol properties. It provides aerosol optical depth at eight spectral channels in the wavelength range of 340–1640 nm. It is a part of the AERONET [41]. The AERONET network provides two types of data: direct Sun spectral data based on the extinction of light through the atmospheric column (Aerosol Optical Depth (AOD), Ångström Exponent (α), SDA (Spectral Deconvolution Algorithm) Fine mode fraction, SDA Fine/Coarse AOD, etc.), and inversion data derived from the angular distribution of the sky radiance (volume particle size distribution, asymmetry factor, complex refractive index, single scattering albedo (SSA), absorption, and extinction optical depths, etc.) [42]. In this study we analyzed the aerosol optical depth (AOD), Ångström Exponent [43], volume particle size distribution, and the single scattering albedo (SSA) from the CLUJ_UBB AERONET station.

2.2. Multi-Wavelength Raman and Depolarization Lidar

Multi-wavelength lidar systems are active remote sensing instruments, which can provide useful information regarding the aerosol properties on different layers due to their high temporal and vertical resolutions. The Cluj-Napoca lidar system (CLOP) emission is based on a Nd-YAG laser Continuum

INLITE II-30, which has a repetition rate of 30 Hz and is equipped with second and third harmonics. The radiation at 1064, 532, and 355 nm is simultaneously emitted into atmosphere. The backscattered radiation is collected by a Cassegrain type telescope (D300 aerosol LIDAR, Raymetrics S.A., Athens, Greece) with a focal length of 1500 mm. The signal detection unit has a total of 6 detection channels, 4 channels for the elastically backscattered radiation at 1064, 532 (cross and parallel), and 355 nm and 2 channels for the Raman radiation backscattered by nitrogen molecules at 607 and 387 nm. The raw data vertical resolution is 3.75 m. The lidar system is part of the EARLINET [44]. For this event the measurements were made during daytime with all the elastic channels and depolarization at 532 nm. Measurement sets of 2 h each were performed with a temporal resolution of 60 s. The calibrated depolarization vertical profiles of the aerosols were computed using the method proposed by [45,46].

2.3. MODIS Retrievals

The Moderate Resolution Imaging Spectroradiometers (MODIS) onboard NASA's Terra and Aqua platforms have been retrieving aerosol parameters since 1999 and 2002, respectively [47]. The polar orbiting satellites perform daily overpasses at 10:30 and 13:30 local solar Equatorial crossing time [48]. Retrieval products are split over land and ocean surfaces by three different sets of algorithms.

The standard Level-2 aerosol product has a 10 km spatial resolution, providing parameters such as total AOD at 550 nm and fine mode fraction based on the spectral fitting error. Other parameters may be derived from lookup tables, including Ångström Exponent, AOD at different wavelengths, and several inversion products. Detailed descriptions of the algorithms are presented extensively in the literature [47–50].

The 3 km product offers the same parameters as the combined land and ocean 10 km algorithm, using a similar structure, inversion methods, and lookup tables. Differences are found in the way reflectance pixels are selected and grouped for retrieval. This higher resolution is particularly useful for tracing fine gradients in smoke and pollution plumes, as well as in dust events. The AOD product selected for this study, Optical_Depth_Land_And_Ocean, provides the highest quality assurance confidence (QAC) for AOD at 0.55 μm, QAC = 3 over land and QAC > 1 over ocean surfaces [51]. Since a similar product for Ångström Exponent is not available at this resolution (3 km), the coarser 10 km product was used as a proxy for estimating aerosol size.

2.4. Modelling Tools

Hybrid Single-Particle Lagrangian Integrated Trajectory model-HYSPLIT back-trajectories were used to analyze the long-range transport of aerosols [52] and to identify the source area.

The NMMB/BSC forecast model developed by Barcelona Supercomputing Center [53] was also used in order to analyze the dynamics of this event.

2.5. In-Situ Measurements

During the analyzed period, in order to identify potential dry deposition of the lofted dust particles, in-situ measurements were performed with a DUSTTRAK™ DRX Aerosol Monitor optical particle counter (Model 8533 with external pump, DUSSTRAK DRX Aerosol Monitor 8533EP, TSI Incorporated, Shoreview, MN, USA) at 1.5 m above ground. It can simultaneously measure size segregated mass fraction concentrations (PM_1, $PM_{2.5}$, Respirable/PM_4, PM_{10}/Thoracic, and Total PM) over a wide concentration range (0.001–150 mg/m^3) in real time.

3. Results and Discussion

3.1. Sun Photometer Data Analysis

For the April the 24th 2019 at Cluj_UBB AERONET site, we analyzed Version 3, level 1.5 data, which are cloud screened and quality controlled, but without the final calibration. As seen in Figure 1a, the Aerosol Optical Depth (AOD) at 500 nm has a daily average of 0.26 ± 0.02, higher than the 0.20

± 0.05 multi-annual monthly average for April at the Cluj_UBB station. Regarding the 440–870 nm Ångström Exponent (Figure 1b), the measured values are low (<1), and specific to coarse aerosols like mineral dust. The volume size distribution (Figure 1c) is dominated by the coarse mode, and the spectral derivative of the single scattering albedo (SSA) is positive in the first part of the measurement interval (Figure 1d, 12:25 UTC), which are consistent with relatively pure mineral dust. The following spectral SSA inversion products (13:26 UTC; 13:45 UTC) exhibit a positive, then negative, spectral derivative, which indicate mixed polluted dust conditions [54,55]. The presence of fine aerosols in the mixture is also indicated by the increase in the fine mode component of the size distribution for 13:26 UTC; 13:45 UTC (Figure 1c).

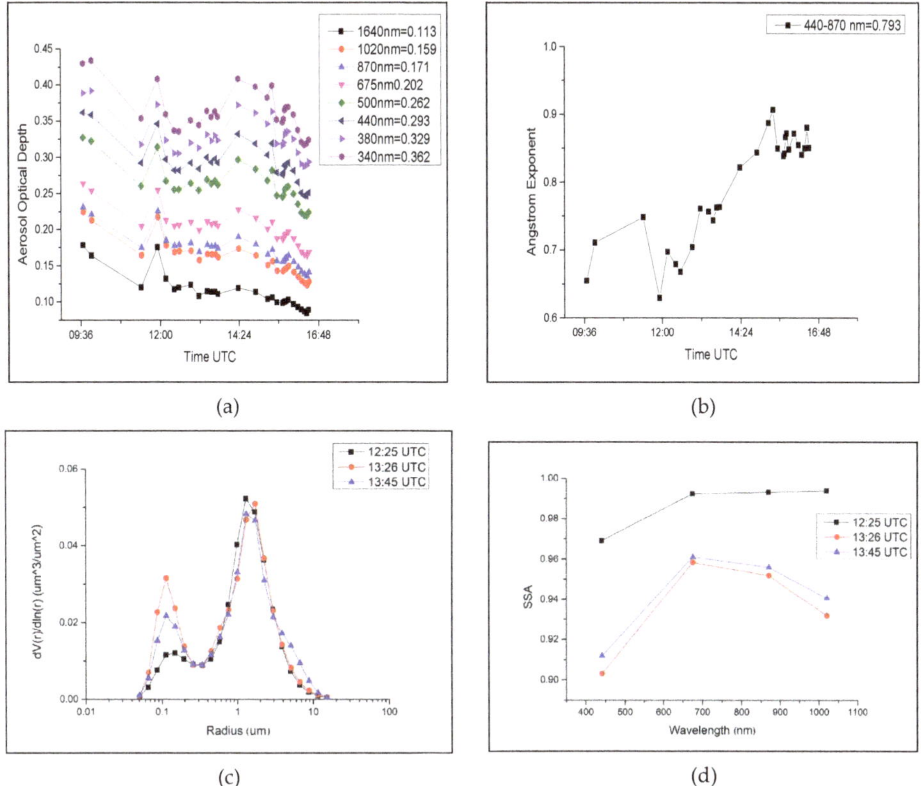

Figure 1. AERONET measurements on 24.04.2019 at CLUJ_UBB station: (**a**) aerosol optical depth (AOD), (**b**) Ångström Exponent, (**c**) Size distribution, (**d**) SSA.

3.2. Lidar Data Analysis

Regarding lidar measurements, in the first set (24.04.2019: 10:15–12:15 UTC), aerosol layers were observed at altitudes between 3200 and 4000 m at 10:15 UTC, and at altitudes from 2500 to 4000 m at 12:15 UTC, as seen in the Range Corrected Signals (RCS) at 1064 (Figure 2a) and 532 nm cross (Figure 2b). At high altitudes, between 8000 and 10,000 m we can identify high cirrus ice clouds, and below 2500 m, two non-dust aerosol layers can be observed.

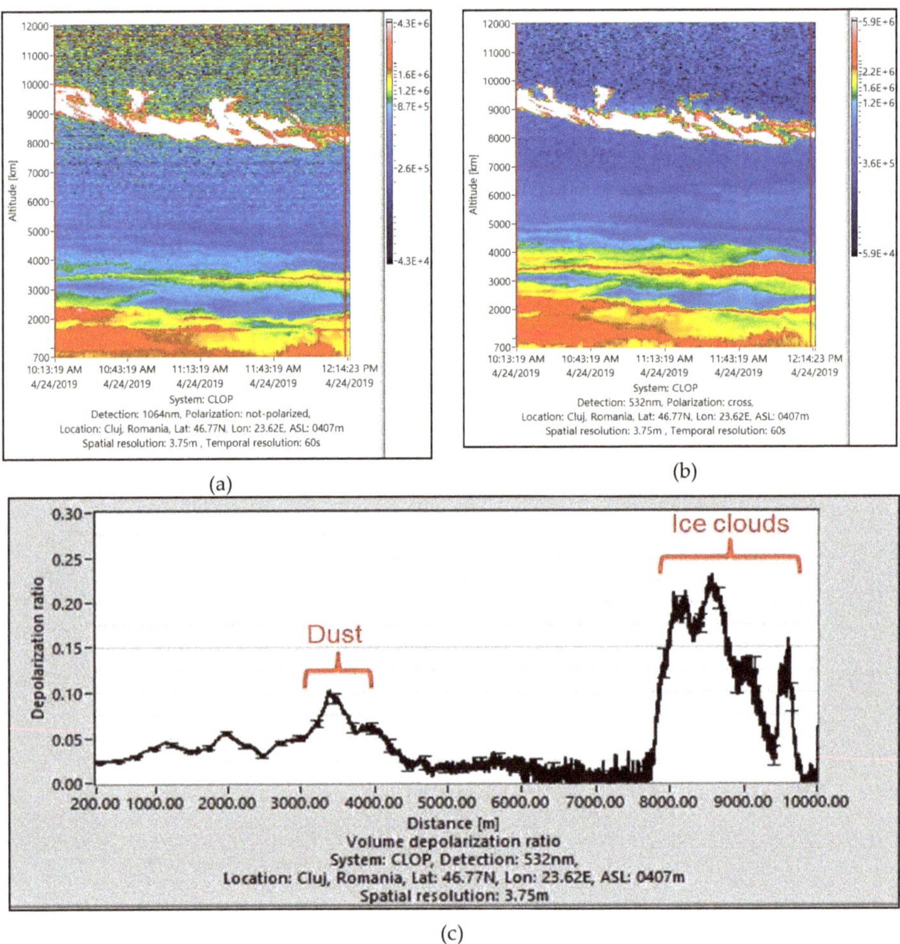

Figure 2. Cluj-Napoca lidar system (CLOP) lidar RCS for 24.04.2019: 10:15–12:15 UTC, Cluj-Napoca, 407 m ASL, at 1064 nm (**a**) and 532 nm cross (**b**); volume depolarization ratio at 532 nm (**c**).

The volume depolarization ratio (Figure 2c) for the higher aerosol layer (2500–4000 m) reaches 0.1, characteristic to Saharan dust [56], while the lower aerosol layers (below 2500 m) exhibit values of 0.05 and 0.06, characteristic to non-dust aerosols (urban-industrial and/or biomass burning aerosols). The particle depolarization ratio for the dust aerosol layer is 0.27, and varies between 0.1 and 0.13 for the lower non-dust aerosol layers, which are in agreement with the values presented in the literature for these aerosol types [57].

In the second lidar measurements set (24.04.2019: 12:20–14:20 UTC) we can see the same aerosol layer between 2500 and 3700 m. In addition, a new layer is detected at higher altitude, at approx. 7000 m, as seen in the RCS at 1064 (Figure 3a) and 532 nm cross (Figure 3b), below the cirrus clouds detected between 7500 and 9500 m. The volume depolarization ratio slightly increased to 0.12, characteristic to Saharan dust (Figure 3c). The particle depolarization ratio for the dust aerosol layer is 0.24, which is in agreement with the values presented in the literature for this aerosol type [57].

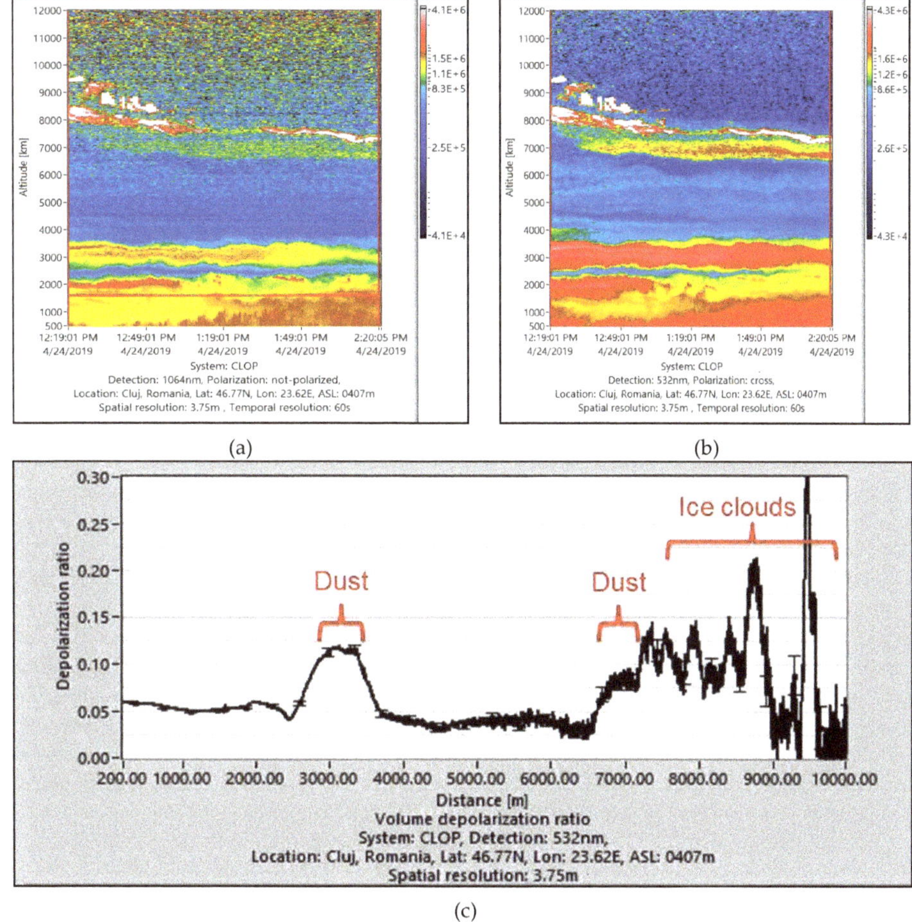

(c)

Figure 3. CLOP lidar RCS for 24.04.2019: 12:20–14:20 UTC, Cluj-Napoca, 407 m ASL, at 1064 nm (**a**) and 532 nm cross (**b**); volume depolarization ratio at 532 nm (**c**).

3.3. MODIS Data Analysis

In Figure 4, AOD values at 550 nm resulting from the 3 × 3 km MODIS level 2 AOD products are represented. Terra MODIS retrieved AOD values between 0.19 and 0.27 over Transylvania, while Aqua MODIS showed AOD values between 0.19 and 0.33. Within the 3 h gap separating the two measurements, an increase of AOD, followed by a decrease of Ångström Exponent values, seemed to indicate higher dust concentrations over the study area. This trend was observed in true color images as air masses traveled East-North East. Collocated lidar measurements (Figure 2) showed thinner layers of dust over Cluj-Napoca at 10:13 UTC, while thicker layers were observed two hours later (Figure 3). Both AOD and Ångström Exponent satellite data showed strong correlation with the equivalent AERONET parameters [58].

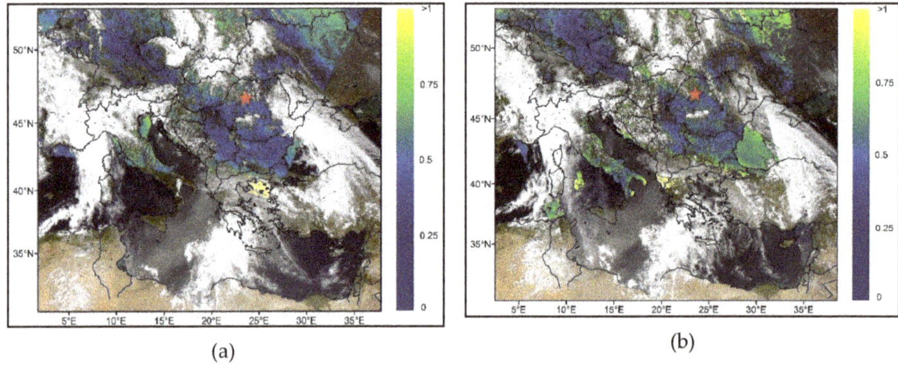

Figure 4. AOD MODIS Terra (**a**) and Aqua (**b**) for April the 24th 2019.

The particle size can be estimated using the Ångström Exponent, as this parameter is inversely related to the particle size. Regarding the range of observed values, because of the relatively low values of Ångström Exponent (less than 1), it is inferred that the majority of particles are coarse. Furthermore, the range of AOD values indicates that the thickness/density of particles in the air does not reach a critical threshold (which leads to an intense decrease in visibility) yet.

In Figure 5 the Ångström Exponent values derived from the 10×10 km MODIS level 2 AOD products are superimposed over a true color image of Eastern Europe. A large mass of dust aerosols can be observed over the Mediterranean Sea and The Balkans. The satellite sensor onboard Terra and Aqua retrieved aerosol parameters at 10:00 UTC and 11:45 UTC, respectively. Terra MODIS retrievals exhibit Ångström Exponent values between 0 and 1.5 over Romania, with lower values ranging from 0 to 1.2 over Transylvania. Aqua MODIS retrievals exhibit Ångström Exponent values between 0 and 1.2 over Romania, with lower values ranging from 0 to 0.95 over Transylvania. These values are consistent with ones reported in the literature for mineral dust [59]. Considering that the dust particle size is usually much bigger in comparison with urban-industrial and biomass burning aerosol, it can be concluded that the analyzed dust plume is not severely polluted nor heavily mixed with smoke.

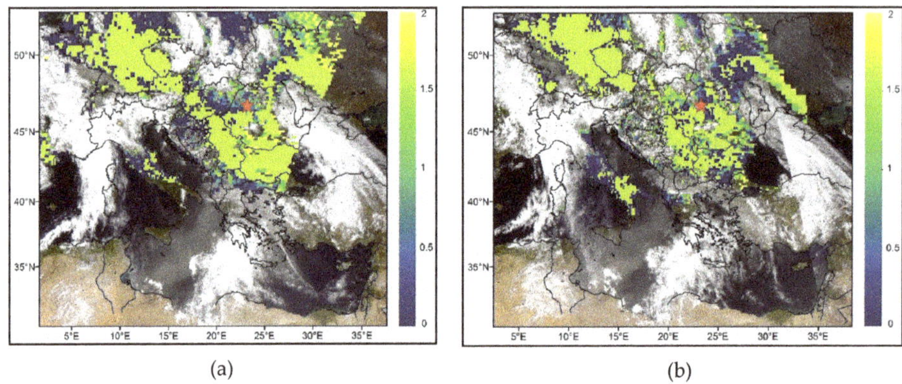

Figure 5. Ångström Exponent MODIS Terra (**a**) and Aqua (**b**) for April the 24th 2019.

3.4. Modelling Data Analysis

The NMMB/BSC dust model [53] predicted the Saharan dust event, as seen in Figure 6. The cyclonic nucleus developed over north-western Africa on April the 20th 2019 (marked with blue arrows in Figure 6a), and lifted up a significant amount of dust that was carried out by upper level southern and south-western circulation (emphasized by blue arrow in Figure 6b). The dust intrusion covered many parts of Europe, including Transylvania. The model predicted a dust load value between 1 and 1.5 g/m^2 over Cluj-Napoca at 12:00 UTC for April the 24th 2019. The 700 hPa wind is the level that corresponds to the layer centered on 3000 m, where the CLOP lidar detected the first dust layer in Figure 3. The south-western circulation over the same area was revealed also at 400 hPa level, corresponding with the altitude of 7000 m (not shown here).

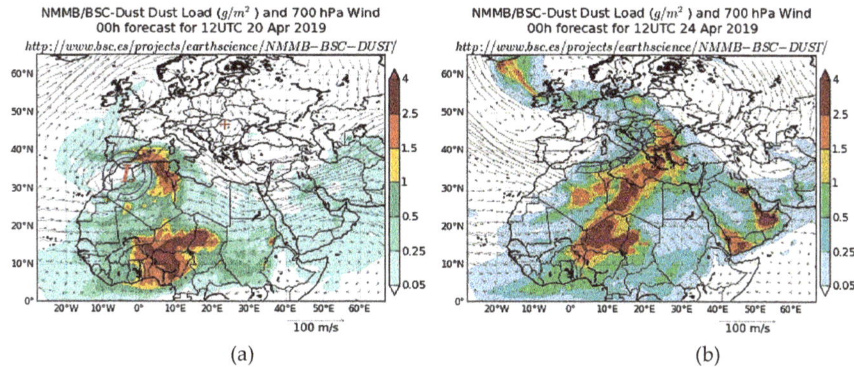

Figure 6. NMMB/BSC model dust load (pallete) and 700 hPa circulation (black arrows) for 20.04.2019, 12 UTC (**a**), and (**b**) for 24.04.2019, 12:00 UTC.

In order to determine the origin of the mineral dust layers detected over Cluj-Napoca, we used the NOAA HYSPLIT back-trajectories model. The back-trajectories (Figure 7) show that the air masses present over Transylvania–Cluj-Napoca station on April the 24th 2019 during the lidar measurements originated from the Saharan desert. Similar to the [60] study, we can analyze the air masses dynamics using Figure 7. For the trajectories that were obtained for a period of 72 h backward in time, the traveling time from the African continent to the detection station (after leaving the African continent) is less than 48 h, which comes in support of the previous assumption that the dust plume is relatively pure.

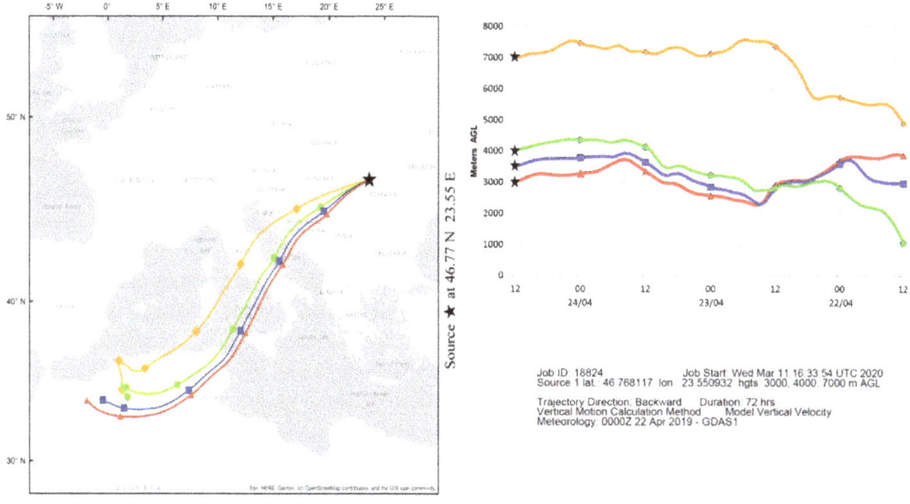

Figure 7. HYSPLIT back-trajectories obtained for 24.04.2019 at 12:00 UTC.

3.5. In-Situ Ground Measurements Analysis

Regarding the in-situ ground level particulate matter values, during the dust intrusion on April the 24th 2019 we measured low values of PM_{10} and $PM_{2.5}$ with an hourly average of 21 μg/m^3 for both PM_{10} and $PM_{2.5}$ and a maximum of 28 μg/m^3 for PM_{10} and 27 μg/m^3 for $PM_{2.5}$. These values are consistent with other measurements reported for Cluj-Napoca city during the period for which data was available from the National Air Quality Network [61]. We can therefore conclude that there was a very limited dry deposition of Saharan dust particles in Cluj-Napoca during this event, and the high-altitude transport continued further north.

4. Conclusions

In this paper, we analyzed a severe dust outbreak over Transylvania, Romania, during April 2019, using multiple remote sensing and in-situ instrumentation supported by modelling tools. These results can also be considered representative for the Pannonian basin.

Results show the presence of dust aerosol layers identified by the multi-wavelength Raman and depolarization lidar at altitudes of 2500–4000 m, and 7000 m, respectively.

An analysis of the optical and microphysical columnar properties from AERONET reveals an increase in the aerosol optical depth (AOD), but the AOD range indicates that the density of lofted dust particles has not reached a critical threshold that would affect visibility. Sun-photometer measurements exhibit low values of the Ångström Exponent (<1) and a dominant coarse mode (>1 μm) in the AERONET particle size distribution.

For the dust layers observed, the volume linear depolarization ratio obtained from the lidar's 532 nm channel is 0.1–0.12, which is characteristic for mineral dust. The 532 nm particle depolarization ratio is 0.24–0.27, and remains relatively constant within the dust layers, indicating that the mineral dust particles are evenly spread.

The variations in the spectral derivative of the SSA, along with lidar information about the presence and depolarization characteristics of non-dust layers below 2500 m indicate the presence of biomass burning and/or urban-industrial aerosols below 2500 m. Therefore, it can be concluded that the dust layers above 2500 m are neither severely polluted nor heavily mixed with smoke.

The analysis of the MODIS retrievals provides added value information on the spatial distribution of the aerosol optical properties (AOD and Ångström Exponent) over Transylvania, and is in agreement with the AERONET ground-based measurements from Cluj-Napoca.

Based on these results, and on the HYSPLIT back-trajectory, NMMB/BSC forecast, and synoptic analysis, we can confirm the presence of lofted Saharan dust layers over Cluj-Napoca, Romania during the April 2019 outbreak.

Collocated in-situ PM monitoring carried out with an optical particle counter showed that dry deposition was low, with PM_{10} and $PM_{2.5}$ concentrations similar to the seasonal averages for Cluj-Napoca.

The case analysis carried out in this paper can be an example of a comprehensive monitoring and characterization of the Saharan dust plume in the north-western part of Romania. The results of this article provide a variety of geo-spatial, temporal and qualitative information on the Saharan dust intrusion in the Transylvania and Pannonian basins. This source of information, obtained from various multi-modal datasets, can help in the forecast and management of this natural hazard in a more efficient way.

Author Contributions: Conceptualization, N.A. and C.B.; Methodology, N.A. and H.Ș.; Visualization, A.R. and A.M., Writing—Review and Editing, N.A., C.B. and H.Ș. All authors have read and agreed to the published version of the manuscript.

Funding: This research was conducted using the research infrastructure purchased within the POSCCE Project entitled "Development of Research Infrastructure for HPC-Based Disaster Management" – MADECIP, SMIS CODE 48801/1862, co-financed by the European Union through the European Regional Development Fund. This work was partially supported by a grant of the Romanian Ministry of Research and Innovation, CCCDI - UEFISCDI, project number 87 PCCDI/2018, project registration code PN-III-P1-1.2-PCCDI-2017-0868, within PNCDI III.

Acknowledgments: The authors gratefully acknowledge the NOAA Air Resources Laboratory (ARL) for the provision of the HYSPLIT transport and dispersion model and READY website (http://www.ready.noaa.gov) used in this publication. Images from the NMMB/BSC-Dust model, operated by the Barcelona Supercomputing Center (http://www.bsc.es/ess/bsc-dust-daily-forecast/) were used in this study. AERONET-Europe/ACTRIS for calibration and maintenance services. The research leading to these results has received funding from European Union's Horizon 2020 research and innovation programme under grant agreement No 654109. The authors would like to acknowledge Simona Andrei, from the National Institute for R&D in Optoelectronics–INOE2000, Magurele, RO for her help in the synoptic analysis.

Conflicts of Interest: The authors declare no conflict of interest.

References

1. Intergovernmental Panel on Climate Change. *Climate Change 2013–The Physical Science Basis: Working Group I Contribution to the Fifth Assessment Report of the Intergovernmental Panel on Climate Change*; Cambridge University Press: Cambridge, UK; New York, NY, USA, 2014. [CrossRef]
2. Díaz, J.; Linares, C.; Carmona, R.; Russo, A.; Ortiz, C.; Salvador, P.; Trigo, R.M. Saharan dust intrusions in Spain: Health impacts and associated synoptic conditions. *Environ. Res.* **2017**, *156*, 455–467. [CrossRef] [PubMed]
3. Stafoggia, M.; Zauli-Sajani, S.; Pey, J.; Samoli, E.; Alessandrini, E.; Basagaña, X.; Cernigliaro, A.; Chiusolo, M.; Demaria, M.; Díaz, J.; et al. MED-PARTICLES Study Group. Desert dust outbreaks in Southern Europe: Contribution to daily PM10 concentrations and short-term associations with mortality and hospital admissions. *Environ. Health Perspect.* **2016**, *124*, 413–419. [CrossRef] [PubMed]
4. Ai, N.; Polenske, K.R. Socioeconomic impact analysis of yellow-dust storms: An approach and case study for Beijing. *Econ. Syst. Res.* **2008**, *20*, 187–203. [CrossRef]
5. Al-Hemoud, A.; Al-Sudairawi, M.; Neelamanai, S.; Naseeb, A.; Behbehani, W. Socioeconomic effect of dust storms in Kuwait. *Arab. J. Geosci.* **2017**, *10*, 18. [CrossRef]
6. Andreae, M. Climate effects of changing atmospheric aerosol levels. In *World Survey of Climatology, Future Climate of the World*; Henderson Sellers, A.H., Ed.; Elsevier: New York, NY, USA, 1995; Volume 16, pp. 341–392.
7. Nicolae, V.; Talianu, C.; Andrei, S.; Antonescu, B.; Ene, D.; Nicolae, D.; Dandocsi, A.; Toader, V.-E.; Ştefan, S.; Savu, T.; et al. Multiyear Typology of Long-Range Transported Aerosols over Europe. *Atmosphere* **2019**, *10*, 482. [CrossRef]
8. Papayannis, A.; Amiridis, V.; Mona, L.; Tsaknakis, G.; Balis, D.; Bösenberg, J.; Chaikovski, A.; De Tomasi, F.; Grigorov, I.; Mattis, I.; et al. Systematic lidar observations of Saharan dust over Europe in the frame of EARLINET (2000–2002). *J. Geophys. Res.* **2008**, *113*. [CrossRef]
9. Salvador, P.; Almeida, S.M.; Cardoso, J.; Almeida-Silva, M.; Nunes, T.; Cerqueira, M.; Alves, C.; Reis, M.A.; Chaves, P.C.; Artinano, B.; et al. Composition and origin of PM10 in Cape Verde: Characterization of long-range transport episodes. *Atmos. Environ.* **2016**, *127*, 326–339. [CrossRef]
10. Ravi, S.; D'Odorico, P.; Breshears, D.D.; Field, J.P.; Goudie, A.S.; Huxman, T.E.; Li, J.; Okin, G.S.; Swap, R.J.; Thomas, A.D.; et al. Aeolian processes and the biosphere. *Rev. Geophys.* **2011**, *49*, RG3001. [CrossRef]
11. Shao, Y.; Wyrwoll, K.-H.; Chappell, A.; Huang, J.; Lin, Z.; Mctainsh, G.H.; Mikami, M.; Tanaka, T.Y.; Wang, X.; Yoon, S. Dust cycle: An emerging core theme in Earth system science. *Aeolian Res.* **2011**, *2*, 181–204. [CrossRef]
12. Querol, X.; Pey, J.; Pandolfi, M.; Alastuey, A.; Cusack, M.; Pérez, N.; Moreno, T.; Viana, M.; Mihalopoulos, N.; Kallos, G.; et al. African dust contributions to mean ambient PM10 mass-levels across the Mediterranean Basin. *Atmos. Environ.* **2009**, *43*, 4266–4277. [CrossRef]
13. Kallos, G.; Astitha, M.; Katsafados, P.; Spyrou, C. Long-Range Transport of Anthropogenically and Naturally Produced Particulate Matter in the Mediterranean and North Atlantic: Current State of Knowledge. *J. Appl. Meteorol. Clim.* **2007**, *46*, 1230–1251. [CrossRef]
14. Lamancusa, C.; Wagstrom, K. Global transport of dust emitted from different regions of the Sahara. *Atmos. Environ.* **2019**, *214*, 1–10. [CrossRef]
15. Sicard, M.; Barragan, R.; Dulac, F.; Alados-Arboledas, L.; Mallet, M. Aerosol optical, microphysical and radiative properties at regional background insular sites in the western Mediterranean. *Atmos. Chem. Phys.* **2016**, *16*, 12177–12203. [CrossRef]

16. Cuevas, E.; Gómez-Peláez, A.J.; Rodríguez, S.; Terradellas, E.; Basart, S.; Garcia, R.D.; Garcia, O.E.; Alonso-Perez, S. The pulsating nature of large-scale Saharan dust transport as a result of interplays between mid-latitude Rossby waves and the North African Dipole Intensity. *Atmos. Environ.* **2017**, *167*, 586–602. [CrossRef]
17. Mona, L.; Amodeo, A.; Pandolfi, M.; Pappalardo, G. Saharan dust intrusions in the Mediterranean area: Three years of Raman lidar measurements. *J. Geophys. Res.* **2006**, *111*, D16. [CrossRef]
18. Marinou, E.; Amiridis, V.; Binietoglou, I.; Tsikerdekis, A.; Solomos, S.; Proestakis, E.; Konsta, D.; Papagiannopoulos, N.; Tsekeri, A.; Vlastou, G.; et al. Three-dimensional evolution of Saharan dust transport towards Europe based on a 9-year EARLINET-optimized CALIPSO dataset. *Atmos. Chem. Phys.* **2017**, *17*, 5893–5919. [CrossRef]
19. Israelevich, P.; Ganor, E.; Alpert, P.; Kishcha, P.; Stupp, A. Predominant transport paths of Saharan dust over the Mediterranean Sea to Europe. *J. Geophys. Res.* **2012**, *117*, 2205. [CrossRef]
20. Janicka, L.; Stachlewska, I.S.; Veselovskii, I.; Baars, H. Temporal variations in optical and microphysical properties of mineral dust and biomass burning aerosol derived from daytime Raman lidar observations over Warsaw, Poland. *Atmos. Environ.* **2017**, *169*, 162–174. [CrossRef]
21. Osborne, M.; Malavelle, F.F.; Adam, M.; Buxmann, J.; Sugier, J.; Marenco, F.; Haywood, J. Saharan dust and biomass burning aerosols during ex-hurricane Ophelia: Observations from the new UK lidar and sun-photometer network. *Atmos. Chem. Phys.* **2019**, *19*, 3557–3578. [CrossRef]
22. Bègue, N.; Tulet, P.; Chaboureau, J.-P.; Roberts, G.; Gomes, L.; Mallet, M. Long-range transport of Saharan dust over northwestern Europe during EUCAARI 2008 campaign: Evolution of dust optical properties by scavenging. *J. Geophys. Res.* **2012**, *117*, D17201. [CrossRef]
23. Abdelkader, M.; Metzger, S.; Mamouri, R.E.; Astitha, M.; Barrie, L.; Levin, Z.; Lelieveld, J. Dust–air pollution dynamics over the eastern Mediterranean. *Atmos. Chem. Phys.* **2015**, *15*, 9173–9189. [CrossRef]
24. Ștefănie, H.; Ajtai, N.; Botezan, C.; Țoancă, F.; Török, Z.; Ozunu, A. Detection of a desert dust intrusion over ClujNapoca, Romania using an elastic backscatter LIDAR system. *Ecoterra-J. Environ. Res. Prot.* **2015**, *12*, 50–55.
25. Tudose, O.G. Contributions to the Study of Atmospheric Aerosols Optical Properties Using Remote Sensing Techniques. Ph.D. Thesis, Alexandru Ioan Cuza University of Iasi, Iasi, Romania, 2013.
26. Rogora, M.; Mosello, R.; Marchetto, A. Long-term trends in the chemistry of atmospheric deposition in Northwestern Italy: The role of increasing Saharan dust deposition. *Tellus B Chem. Phys. Meteorol.* **2004**, *56*, 426–434. [CrossRef]
27. Papagiannopoulos, N.; D'Amico, G.; Gialitaki, A.; Ajtai, N.; Alados-Arboledas, L.; Amodeo, A.; Amiridis, V.; Baars, H.; Balis, D.; Binietoglou, I.; et al. An EARLINET Early Warning System for atmospheric aerosol aviation hazards. *Atmos. Chem. Phys. Discuss.* **2020**. in review. [CrossRef]
28. Cazacu, M.M.; Tudose, O.G.; Timofte, A.; Rusu, O.; Apostol, L.; Leontie, L.; Gurlui, S. A case study of the behavior of aerosol optical properties under the incidence of a saharan dust intrusion event. *Appl. Ecol. Environ. Res.* **2016**, *14*, 183–194. [CrossRef]
29. Labzovskii, L.; Toanca, F.; Nicolae, D. Determination of saharan dust properties over bucharest, Romania part 2: Study cases analysis. *Rom. J. Phys.* **2014**, *59*, 1097–1108.
30. Gothard, M.; Nemuc, A.; Radu, C.; Dascalu, S. An intensive case of saharan dust intrusion over south east Romania. *Rom. Rep. Phys.* **2014**, *66*, 509–519.
31. Rosu, I.A.; Cazacu, M.M.; Prelipceanu, O.S.; Agop, M. A Turbulence-Oriented Approach to Retrieve Various Atmospheric Parameters Using Advanced Lidar Data Processing Techniques. *Atmosphere* **2019**, *10*, 38. [CrossRef]
32. Ajtai, N.; Stefanie, H.; Arghius, V.; Meltzer, M.; Costin, D. Characterization of Aerosol Optical and Microphysical Properties Over North-Western Romania In Correlation with Predominant Atmospheric Circulation Patterns. In Proceedings of the 17th International Multidisciplinary Scientific Geoconference (SGEM 2017), Albena, Bulgaria, 29 June–5 July 2017. [CrossRef]
33. Ajtai, N.; Ștefănie, H.; Ozunu, A. Description of aerosol properties over Cluj-Napoca derived from AERONET sun photometric data. *Environ. Eng. Manag. J.* **2013**, *12*, 227–232. [CrossRef]
34. Talianu, C.; Nemuc, A.; Nicolae, D.; Cristescu, C.P. Dust Event Detection from Lidar Measurements. *Sci. Bull. Politeh. Univ. Buchar. Ser. A Appl. Math. Phys.* **2007**, *69*, 53–62.

35. Cazacu, M.M.; Tudose, O.; Boscornea, A.; Buzdugan, L.; Timofte, A.; Nicolae, D. Vertical and temporal variation of aerosol mass concentration at Magurele–Romania during EMEP/PEGASOS campaign. *Rom. Rep. Phys.* **2017**, *69*, 1–15.

36. Mărmureanu, L.; Marin, C.A.; Andrei, S.; Antonescu, B.; Ene, D.; Boldeanu, M.; Vasilescu, J.; Vițelaru, C.; Cadar, O.; Levei, E. Orange Snow—A Saharan Dust Intrusion over Romania During Winter Conditions. *Remote Sens.* **2019**, *11*, 2466. [CrossRef]

37. Severe Weather Europe. Available online: http://www.severe-weather.eu/mcd/extent-of-airborne-saharan-dust-over-europe-on-wednesday-and-thursday/ (accessed on 3 October 2019).

38. Pappalardo, G.; Amodeo, A.; Apituley, A.; Comeron, A.; Freudenthaler, V.; Linné, H.; Ansmann, A.; Bösenberg, J.; D'Amico, G.; Mattis, I.; et al. EARLINET: Towards an advanced sustainable 20 European aerosol lidar network. *Atmos. Meas. Tech.* **2014**, *7*, 2389–2409. [CrossRef]

39. Holben, B.N.; Eck, T.F.; Slutsker, I.; Tanre, D.; Buis, J.P.; Setzer, A.; Vermote, E.; Reagan, J.A.; Kaufman, Y.; Nakajima, T.; et al. AERONET-A federated instrument network and data archive for aerosol characterization. *Remote Sens. Environ.* **1998**, *66*, 1–16. [CrossRef]

40. ACTRIS. Available online: https://www.actris.eu/ (accessed on 6 September 2019).

41. AERONET. Available online: https://aeronet.gsfc.nasa.gov/ (accessed on 12 September 2019).

42. Dubovik, O.; King, M. A flexible inversion algorithm for retrieval of aerosol optical properties from Sun and sky radiance measurements. *J. Geophys. Res.* **2000**, *105*, 20673–20696. [CrossRef]

43. Ångström, A. The parameters of atmospheric turbidity. *Tellus* **1964**, *16*, 64–75. [CrossRef]

44. European Aerosol Research Lidar Network. Available online: https://www.earlinet.org (accessed on 18 September 2019).

45. Freudenthaler, V. About the efects of polarising optics on lidar signals and the D90 calibration. *Atmos. Meas. Tech.* **2016**, *9*, 4181–4255. [CrossRef]

46. Belegante, L.; Bravo-Aranda, J.A.; Freudenthaler, V.; Nicolae, D.; Nemuc, A.; Ene, D.; Alados-Arboledas, L.; Amodeo, A.; Pappalardo, G.; D'Amico, G.; et al. Experimental techniques for the calibration of lidar depolarization channels in EARLINET. *Atmos. Meas. Tech.* **2018**, *11*, 1119–1141. [CrossRef]

47. Sayer, A.M.; Munchak, L.A.; Hsu, N.C.; Levy, R.C.; Bettenhausen, C.; Jeong, M.-J. MODIS Collection 6 aerosol products: Comparison between Aqua's e-Deep Blue, Dark Target, and merged data sets, and usage recommendations. *J. Geophys. Res.* **2014**, *119*, 13965–13989. [CrossRef]

48. Levy, R.C.; Remer, L.A.; Tanré, D.; Mattoo, S.; Kaufman, Y.J. Algorithm for Remote Sensing of Tropospheric Aerosol over Dark Targets from MODIS: Collections 005 and 051: Revision 2; MODIS Algorithm Theoretical Basis Document. 2009. Available online: https://atmosphere-imager.gsfc.nasa.gov/sites/default/files/ModAtmo/ATBD_MOD04_C005_rev2_0.pdf (accessed on 3 October 2019).

49. Levy, R.C.; Mattoo, S.; Munchak, L.A.; Remer, L.A.; Sayer, A.M.; Patadia, F.; Hsu, N.C. The Collection 6 MODIS aerosol products over land and ocean. *Atmos. Meas. Tech.* **2013**, *6*, 2989–3034. [CrossRef]

50. Hsu, N.C.; Jeong, M.-J.; Bettenhausen, C.; Sayer, A.M.; Hansell, R.; Seftor, C.S.; Huang, J.; Tsay, S.-C. Enhanced Deep Blue aerosol retrieval algorithm: The second generation. *J. Geophys. Res. Atmos.* **2013**, *118*, 9296–9315. [CrossRef]

51. Remer, L.A.; Mattoo, S.; Levy, R.C.; Munchak, L.A. MODIS 3 km aerosol product: Algorithm and global perspective. *Atmos. Meas. Tech.* **2013**, *6*, 1829–1844. [CrossRef]

52. Draxler, R.R.; Rolph, G.D. HYSPLIT (HYbrid Single-Particle Lagrangian Integrated Trajectory) Model Access via NOAA ARL READY Website. NOAA Air Resources Laboratory. 2015; College Park. Available online: http://www.arl.noaa.gov/HYSPLIT.php (accessed on 11 March 2020).

53. Pérez, C.; Haustein, K.; Janjic, Z.; Jorba, O.; Huneeus, N.; Baldasano, J.M.; Black, T.; Basart, S.; Nickovic, S.; Miller, R.L.; et al. An online mineral dust aerosol model for meso to global scales: Model description, annual simulations and evaluation. *Atmos. Chem. Phys.* **2011**, *11*, 13001–13027. [CrossRef]

54. Dubovik, O.; Holben, B.; Eck, T.; Smirnov, A.; Kaufman, Y.; King, M.; Tanré, D.; Slutsker, I. Variability of Absorption and Optical Properties of Key Aerosol Types Observed in Worldwide Locations. *J. Atmos. Sci.* **2002**, *59*, 590–608. [CrossRef]

55. Li, J.; Carlson, B.E.; Lacis, A.A. Using single-scattering albedo spectral curvature to characterize East Asian aerosol mixtures. *J. Geophys. Res. Atmos.* **2015**, *120*, 2037–2052. [CrossRef]

56. Groß, S.; Gasteiger, J.; Freudenthaler, V.; Wiegner, M.; Geiß, A.; Schladitz, A.; Toledano, C.; Kandler, K.; Tesche, M.; Ansmann, A.; et al. Characterization of the planetary boundary layer during SAMUM-2 by means of lidar measurements. *Tellus B* **2011**, *63*, 695–705. [CrossRef]

57. Tesche, M.; Gross, S.; Ansmann, A.; Müller, D.; Althausen, D.; Freudenthaler, V.; Esselborn, M. Profiling of Saharan dust and biomass-burning smoke with multiwavelength polarization Raman lidar at Cape Verde. *Tellus B Chem. Phys. Meteorol.* **2011**, *63*, 649–676. [CrossRef]

58. Li, L.; Sokolik, I. Analysis of Dust Aerosol Retrievals Using Satellite Data in Central Asia. *Atmosphere* **2018**, *9*, 288. [CrossRef]

59. Kalashnikova, O.V.; Kahn, R.A. Mineral dust plume evolution over the Atlantic from MISR and MODIS aerosol retrievals. *J. Geophys. Res.* **2008**, *113*, D24. [CrossRef]

60. Soupiona, O.; Samaras, S.; Ortiz-Amezcua, P.; Böckmann, C.; Papayannis, A.; Moreira, G.A.; Benavent-Oltra, J.A.; Guerrero-Rascado, J.L.; Bedoya-Velasquez, A.E.; Olmo, F.J.; et al. Retrieval of optical and microphysical properties of transported Saharan dust over Athens and Granada based on multi-wavelength Raman lidar measurements: Study of the mixing processes. *Atmos. Environ.* **2019**, *214*, 116824. [CrossRef]

61. Calitate Aer. Available online: www.calitateaer.ro (accessed on 25 September 2019).

Article

Cloud Occurrence Frequency at Puy de Dôme (France) Deduced from an Automatic Camera Image Analysis: Method, Validation, and Comparisons with Larger Scale Parameters

Jean-Luc Baray [1,2,*], Asmaou Bah [1], Philippe Cacault [2], Karine Sellegri [1], Jean-Marc Pichon [1,2], Laurent Deguillaume [1,2], Nadège Montoux [1], Vincent Noel [3], Geneviève Seze [4], Franck Gabarrot [5], Guillaume Payen [5] and Valentin Duflot [5,6]

[1] Laboratoire de Météorologie Physique, UMR 6016, CNRS, Université Clermont Auvergne, 63178 Aubière, France; b.asmaoubah@gmail.com (A.B.); K.Sellegri@opgc.univ-bpclermont.fr (K.S.); j.m.pichon@opgc.fr (J.-M.P.); l.deguillaume@opgc.univ-bpclermont.fr (L.D.); n.montoux@opgc.univ-bpclermont.fr (N.M.)

[2] Observatoire de Physique du Globe de Clermont-Ferrand, UMS 833, CNRS, Université Clermont Auvergne, 63178 Aubière, France; P.Cacault@opgc.univ-bpclermont.fr

[3] Laboratoire d'Aérologie, UMR5560, CNRS, Université Paul Sabatier Toulouse III, 31400 Toulouse, France; vincent.noel@aero.obs-mip.fr

[4] Laboratoire de Météorologie Dynamique, UMR8539, CNRS, ENS, Ecole polytechnique, Université Pierre Marie Curie, 75252 Paris, France; genevieve.seze@lmd.jussieu.fr

[5] Observatoire OSU-Réunion, UMS3365, CNRS, Université de la Réunion, 97444 Saint Denis de la Réunion, France; franck.gabarrot@univ-reunion.fr (F.G.); guillaume.payen@univ-reunion.fr (G.P.); valentin.duflot@univ-reunion.fr (V.D.)

[6] Laboratoire de l'Atmosphère et des Cyclones, UMR8105, CNRS, Météo-France, Université de la Réunion, 97444 Saint Denis de la Réunion, France

[*] Correspondence: J.L.Baray@opgc.fr; Tel.: +33-473405133

Received: 31 October 2019; Accepted: 10 December 2019; Published: 13 December 2019

Abstract: We present a simple algorithm that calculates the cloud occurrence frequency at an altitude site using automatic camera image analysis. This algorithm was applied at the puy de Dôme station (PUY, 1465 m. a.s.l., France) over 2013–2018. Cloud detection thresholds were determined by direct comparison with simultaneous in situ cloud probe measurements (particulate volume monitor (PVM) Gerber). The cloud occurrence frequency has a seasonal cycle, with higher values in winter (60%) compared to summer (24%). A cloud diurnal cycle is observed only in summer. Comparisons with the larger scale products from satellites and global model reanalysis are also presented. The NASA cloud-aerosol transport system (CATS) cloud fraction shows the same seasonal and diurnal variations and is, on average, 11% higher. Monthly variations of the ECMWF ERA-5 fraction of cloud cover are also highly correlated with the camera cloud occurrence frequency, but the values are 19% lower and up to 40% for some winter months. The METEOSAT-SEVIRI cloud occurrence frequency also follows the same seasonal cycle but with a much smaller decrease in summer. The all-sky imager cloud fraction (CF) presents larger variability than the camera cloud occurrence but also follows similar seasonal variations (67% in winter and 44% in summer). This automatic low-cost detection of cloud occurrence is of interest in characterizing altitude observation sites, especially those that are not yet equipped with microphysical instruments and can be deployed to other high-altitude sites equipped with cameras.

Keywords: cloud occurrence frequency; automatic camera image analysis

1. Introduction

Clouds can act as a greenhouse ingredient to warm the Earth by trapping outgoing longwave radiation and can also cool the earth by reflecting shortwave solar radiation [1]. The net effect of these two competing processes depends on the height, type, and optical properties of the clouds. Clouds also play a major role in the hydrological cycle, acting as the connection between atmospheric water vapor and rain [2], and are involved in climate feedback processes [3,4]. Cloud microphysical properties are an important characteristic of clouds, as their ability to produce rain and snow, generate lightning, and contribute to the radiation balance of the earth depends on local air motions but also on their individual microphysical properties [5]. The chemical composition of clouds is also an important issue. The chemical and biological transformations occurring in clouds perturb both the gas phase chemical composition and the microphysical and chemical properties of aerosols acting as cloud condensation nuclei (CCN). These transformations could, therefore, impact air quality and cloud formation. For example, a recent study showed that cloud microorganisms isolated at the puy de Dôme station (PUY) were able to produce biosurfactants [6], which could favor the conversion of aerosol into cloud droplets, based on Köhler's theory [7].

In situ techniques for clouds' microphysical and chemical characterizations are currently deployed at altitude sites that are privileged sites for studying clouds during specific measurement campaigns or continuous observations. The monitoring of cloud properties from the situ measurements at high-altitude sites over long term periods is one of the goals of the ACTRIS research infrastructure [8], in order to assess multi-decanal tendencies in cloud properties over Europe. In pursuit of a harmonized European cloud in situ data set, an inter-comparison campaign, performed at PUY in May 2013, gathered a unique set of cloud instruments, which are operated simultaneously in ambient air conditions and in a wind tunnel, for assessing the inter-instrumental variability of microphysical cloud measurements. The study highlighted the good general agreement in the sizing abilities of the different instruments but also some discrepancies in their ability to assess the amplitude and variability of the cloud droplet number concentration [9]. Long-term chemical and biological composition are also investigated [10–12], with the objective to classify clouds based on their chemical properties and analyze their backward trajectories. Specific field campaigns are regularly organized, where clouds are sampled and analyzed [13,14].

Apart from the detailed cloud microphysical structure, which provides key information for evaluating the impact of clouds on climate, information on whether a cloud is present with good temporal resolution is already valuable for an atmospheric observation station. For example, this information can be useful for:

-   The validation of cloud sensors aboard satellite missions;
-   The organization of ongoing or future field measurement campaigns. The statistical analysis of cloud presence depending on the days and seasons is valuable information for choosing a suitable date for a field campaign focused on cloud studies or, on the contrary, to avoid the presence of clouds for field campaigns that require clear sky conditions;
-   The constitution of sets of metadata to analyze other observations (gas, aerosol, etc.).
-   Knowledge of cloud occurrence thus offers important auxiliary data for altitude observation sites, which must be established with reliable instruments and methods. Probes, such as the particulate volume monitor (PVM) Gerber, can provide this information, but there may be problems with the reliability or continuity of the data for a long-term survey due to icing conditions. More generally, not all sites are equipped with microphysical probes, which are expensive and need assistance under harsh meteorological conditions, such as intensive icing.

A low-cost alternative to cloud microphysical probes is presented here for the detection of clouds at and below a high-altitude site. The objectives of this article are to demonstrate the feasibility and reliability of the automatic detection of clouds from camera images on the PUY site, to present comparisons of this new product with other data, and to provide the first cloud climatology study for the site.

2. Instruments and Data

2.1. Network and All-Sky Cameras

Cézeaux and PUY are two instrumented sites of CO-PDD (Cézeaux-Aulnat-Opme-Puy de Dôme), a full instrumented platform for atmospheric research, which is nationally labeled by CNRS, the French national center for scientific research, and has been recognized as a global GAW (global atmospheric watch, station PUY) station since 2014 [15].

Two network cameras are in operation—one at the Cézeaux site looking towards PUY, and the other at PUY looking towards the Clermont-Ferrand area. Cézeaux is a campus at the University Clermont Auvergne, located south of the Clermont-Ferrand city, at an altitude of 410 m a.s.l. PUY is the mountain station, located 12 km east of Cézeaux, at an altitude 1465 m a.s.l. (Figure 1).

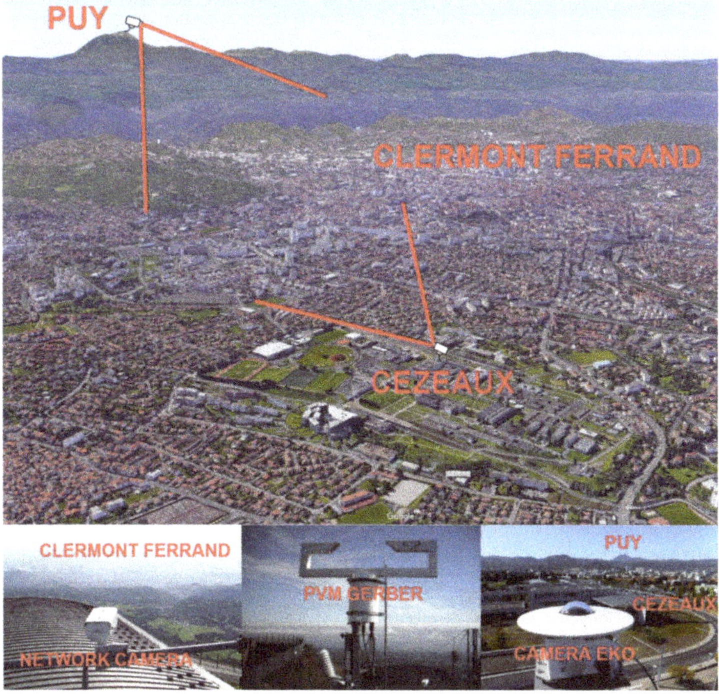

Figure 1. Geographical configuration and fields of view of the network cameras in operation at Cézeaux and puy de Dôme station (PUY) sites superimposed to a ©Google Map image (**top**) and photographs of the PUY platform hosting the network camera (**bottom left**), PVM Gerber (**bottom center**), and the EKO all-sky imager on the Cézeaux site (**bottom right**). A camera similar to the one in operation at PUY is also in operation of the Cézeaux site (not used in this work) but is not visible in the photo.

The two network cameras are Axis P1343 cameras, which are widely used in daytime and nighttime video surveillance. These cameras capture 600 × 800 pixel images every ten minutes in an uncompressed JPEG format. More technical details on the cameras are available on the Axis manufacturer's website [16].

The images taken by the Cézeaux and PUY cameras are available in real time on the website of the Observatoire de Physique du Globe de Clermont-Ferrand [17]. The percentage of available network camera images related to the maximum number of images theoretically available varied from 72% to 100%, except for May 2018 (55%, Table 1). Images were missed due to problems while writing the images, data transmission, electrical shutdown of the station, or accumulation of snow in front of the camera during winter.

In addition, an all-sky camera has been in operation at Cézeaux since December 2015. The all-sky camera is an EKO SRF-02, a fully automatic imaging system that captures 1024×768 pixel JPEG images of the total sky at different exposures every 2 min. When the sky images are captured, a real time cloud fraction (CF) is automatically calculated by the acquisition software, which offers great flexibility and functions. For example, the software is able to define the area of interest by horizon masking. A more optimized CF evaluation was also performed by the software ELIFAN [18], which is used in this study.

The SRF-02 is connected to a computer network through a standard Ethernet interface. The all-sky camera is equipped with an air pump and drying cartridge. In this paper, the methodology is based on camera images from the PUY camera, and PVM Gerber has been used for validation. In this work, 524 days of operation of the all-sky cameras have been used from December 2015 to December 2018.

Table 1. The percentage of available network camera images at PUY related to the maximum number of images theoretically available per month (lines) and year (columns).

Month	2013	2014	2015	2016	2017	2018
January	99.3%	87.6%	90.6%	93.9%	94.0%	100%
February	100%	99.3%	76.1%	99.9%	100%	95.1%
March	92.1%	99.4%	95.8%	98.7%	99.0%	99.9%
April	98.7%	100%	97.8%	97.6%	100%	97.0%
May	92.9%	99.7%	92.5%	94.9%	99.9%	55.5%
June	98.6%	98.5%	88.2%	90.9%	100%	99.7%
July	97.4%	97.4%	96.2%	92.3%	96.8%	96.7%
August	95.8%	99.7%	86.1%	96.6%	96.3%	94.1%
September	98.3%	96.3%	92.8%	98.7%	99.6%	97.6%
October	96.0%	98.6%	96.4%	99.5%	98.1%	82.9%
November	92.2%	95.1%	100%	95.8%	100%	100%
December	88.7%	72.5%	97.1%	89.9%	100%	98.0%

2.2. PVM Gerber

Gerber PVM-100 is a ground-direct scattering laser spectrophotometer for cloud droplet volume measurements and is manufactured by Gerber Scientific, Inc., Reston, Virginia (USA). It measures the cloud liquid water content (LWC), the surface of the droplets, and determines the effective radius of the droplets [19,20].

A laser light emitted at a wavelength of $\lambda = 0.780$ μm is dispersed by the cloud droplets passing through the sampling volume of the 3 cm^{-3} probe. This light is collected by a lens system whose angle varies from 0.25° to 5.2°. The scattered light is converted into a signal that is proportional to the droplet density (or LWC) and the particle surface density (PSA) (Gerber et al., 1994). This light measures the cloud water content (in the range of 0.002–10 g m^{-3}), the total particle area, and, deduced from these values, the effective droplet radius (in the range of 2–70 μm).

In this paper, we use the LWC measured with the PVM cloud probe as a reference to establish the presence of clouds at the station. If the LWC > 0 g m^{-3}, we consider the station to be under cloudy conditions. Otherwise, if LWC $= 0$ g m^{-3}, the summit is considered to be not cloudy.

2.3. CATS (Cloud-Aerosol Transport System)

The NASA cloud-aerosol transport system (CATS) is a lidar onboard the International Space Station (ISS) that provides vertical profiles of atmospheric aerosols and clouds [21]. CATS was in operation from February 2015 to October 2017.

The CF parameter is deduced from the calibrated vertical profiles of the backscattered signals at 1064 nm [22], with an algorithm separating low level clouds and aerosols similar to the cloud-aerosol lidar and infrared pathfinder satellite observation (CALIPSO) algorithm [23]. More details on the CATS inversion algorithms and data can be found in the CATS data product catalog [24].

In the present work, we have used cloud detections from the CATS level 2 operation cloud layer product (L2O-CLay). This product includes cloud layers detected horizontally and averaging between 1 to 60 km, reported every 5 km along-track. To document the daily variability of the cloud detections in the present study, we extracted all profiles measured by CATS in a $5° \times 5°$ latitude-longitude box centered on PUY, with a feature type score above 5 (as in [23]), and indexed them according to the local time when the observation was made.

2.4. Meteosat/SEVIRI

The SEVIRI (spinning enhanced visible and infrared imager) onboard MSG (METEOSAT second generation) geostationary satellite is a visible and infrared multi-channel imager that operates on a 15 min repeat cycle. The spatial resolution at a sub-satellite point is 3×3 km^2 and about 3.2×5.3 km^2 close to PUY. The presence of clouds can be detected for each elementary pixel, and the cloud occurrence can thus be established.

Cloud detection is also divided into types in the vertical dimension. The cloud type classification used here is obtained from the satellite application facility for NoWCasting (SAF NWC) algorithm [25], developed by Derrien and Le Gleau [26,27]. In the SAF NWC algorithm, cloud detection and classification rely on multi-spectral threshold tests applied at the pixel scale to a set of spectral and textural features. Thresholds are values that depend on illumination and viewing geometry, as well as geographical location. These values are computed using ancillary data and radiative transfer models.

After sorting between cloud free and cloud contaminated pixels, cloudy pixels are classified into ten sets, according to the altitude of the cloud cover summit. For high-topped clouds, a sub-division is performed according to their semi-transparency.

In this study, we consolidate these ten classes into five main classes: low-topped clouds, mid-topped clouds, opaque high-topped clouds, non-opaque high-topped clouds, and partial cloud cover. For pixels belonging to the "partial cloud class", no pressure information is available. Clouds partly covering these pixels can be at any level. Recently, this cloud type classification has been evaluated using CALIOP lidar data [28]. High-topped clouds, including thin cirrus clouds, are well observed. Low-topped clouds over land are less well observed when they are thin or only partially cover the pixels or during nigh-time when they are close to the ground.

2.5. ECMWF Reanalysis

ERA-5 is a recent global atmospheric reanalysis tool produced by ECMWF, which has available since 2018 and succeeded ERA-Interim. ERA-5 contains several improvements to ERA-Interim: the use of more recent versions of data assimilation methods and parameterizations of Earth processes, as well as improved temporal and spatial resolutions from 6 h in ERA-Interim to hourly in ERA-5, and from 79 km in the horizontal dimension and 60 levels in the vertical in ERA-Interim to 31 km and 137 levels in ERA-5 [29].

In this work, we extract the parameter "fraction of cloud cover" from the ECMWF ERA-5 archive at the 850 pressure level from 1 January 2013 to 31 December 2018 at an hourly temporal resolution and a 0.125° spatial resolution and linearly interpolate at the exact location of PUY (45.77° N, 2.95° E). This parameter is the proportion of a grid box covered by cloud (liquid or ice) and varies between zero and one.

Notably, this parameter is comparable to the CF established with the all-sky camera, but it is different from the cloud occurrence established based on the temporal evolution of the cloud or no cloud binary parameter with the network camera or Meteosat SEVIRI. In this work, we will thus use the term cloud fraction (CF in abbreviated form) when the cloud cover is established spatially and "cloud occurrence frequency" (COF in abbreviated form) when it is established temporally.

3. Methodology of Network Camera Image Processing

The camera produces images in the joint photographic experts group (JPEG) format. A JPEG image is a matrix of pixels ranging from the value 0 (for low light) to 255 (for high brightness) in 3 channels for red green and blue colors. It is possible to access and manipulate the digital accounts of the images' pixels with scientific software. We used the commercial MATLAB software environment [30], but the code can easily be adapted for open access software, such as Python [31].

The methodology is based on the detection of contrast in specific areas of the image. The detection of contrast is based on the standard deviation of the pixel values on an area of the image. When the camera is in a cloud, the image has little contrast, and the standard deviation value is low. On the contrary, for a non-cloudy situation, an area centered on Clermont-Ferrand will have high contrast corresponding to a high value of standard deviation.

This camera works similarly with any of the three colors; in this work we chose red. This method works during the day, but also at night thanks to night lighting in the city.

A second zone centered on the horizon of the Livradois-Forez Mountains makes it possible to distinguish individual clouds in situations where a sea of clouds is present: low clouds have an altitude of the top of the cloud layer less than the altitude of PUY (1465 m). The sizes of the zones are 100×350 pixels (Clermont-Ferrand) and 50×270 pixels (Livradois-Forez).

The camera may move slightly because of strong wind or maintenance. It is then necessary to calculate the location of the zones relative to a mark on the image to correct possible movements of the camera. We also had to remove incorrect images with rain drops or snow on the lens of the camera. Figure 2 shows some examples of these images and MATLAB processing in cloudy, sea of clouds, and clear sky conditions. Finally, it is necessary to define a threshold value for the standard deviation, which separates cloudy and clear sky situations.

Figure 3 presents a sensitivity analysis applied to the total set of images (2013–2018). The maximum agreement is obtained with a threshold value of 6 for the two zones (Clermont and Livradois-Forez). Agreement is obtained by dividing the number of cases where the two systems (camera and PVM) obtain the same result (cloudy or clear sky) by the total number of coincidences (the time difference between the image taking and the PVM measurement is less than 2 min). The detection of clouds depends on the signal to noise ratio of the camera, which may be different during the day and night. In order to check the robustness of the method during day and night, we have also applied a sensitivity analysis only for daytime (Figure 3b) and nighttime (Figure 3c) images. The results show that a threshold value of 6 correctly separates cloudy and clear sky situations, as well as day and night (between 85% and 90% agreement). Since the threshold value of 6 separates a cloudy situation from a non-cloudy situation, we consider that below 5.5 the image will be considered cloudy, above 6.5 it will be considered cloudless, and between these two values it will be considered indeterminate.

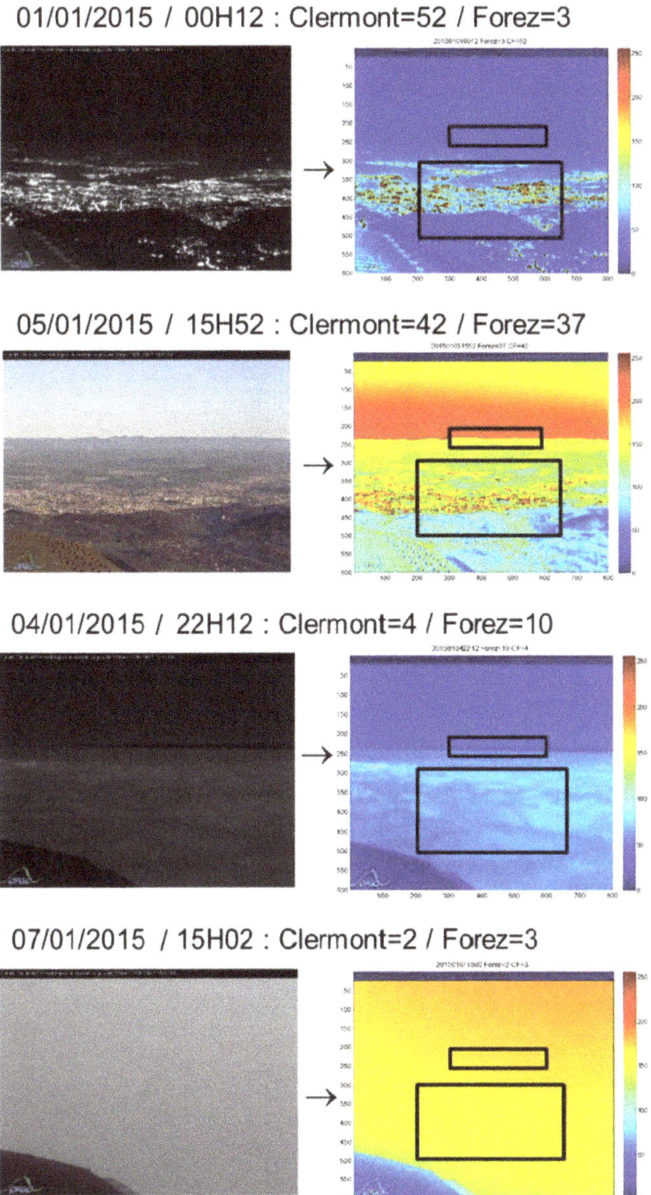

Figure 2. Examples of images and MATLAB processing outputs for different situations, from top to bottom: night time clear sky, day-time clear sky, night-time sea of clouds, and day-time cloud. The two zones are indicated with black rectangles, and the days and hours and standard deviation values in the two areas of Clermont (bottom large rectangle) and Livradois-Forez (top little rectangle) are given on the top of each image.

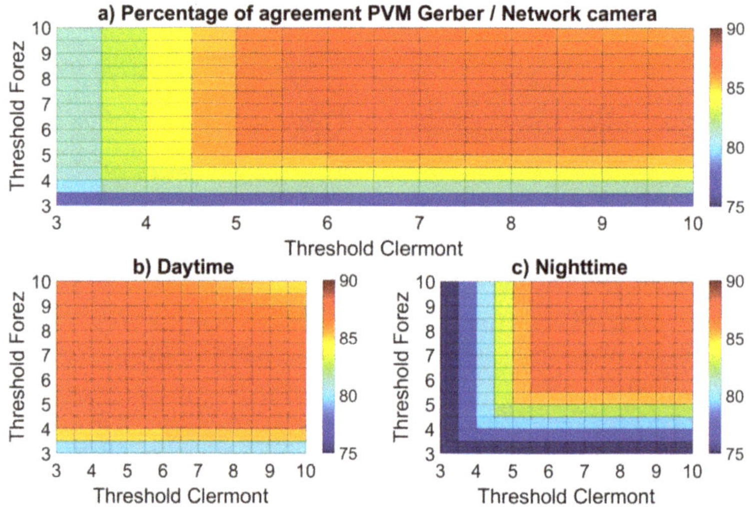

Figure 3. (**a**) Percentage of the agreement between cloud occurrence from the particulate volume monitor (PVM) and camera depending on the threshold values of the camera data processing (Clermont on the X axis, Livradois Forez on the y axis). (**b**) The same plot reduced for daytime images (9 to 16 UT). (**c**) Same plot reduced for nighttime images (23 to 4 UT).

4. Validation

The COF monthly average was calculated over the period 2013–2018 by dividing the number of images in which a cloud is detected, following the method described in Section 3, by the total number of recorded images. For validation purposes, this parameter is compared to the COF deduced from the PVM Gerber measurements. The results are shown in Figure 4. Figure 4a shows a very good agreement between the two techniques over some periods. Figure 4b shows that the months during which a large difference is found between the two technics correspond to a very low operating rate of the PVM Gerber by taking all the measurements available, not only the measurements coincident with the camera.

Table 2 provides the correlation coefficients between the two series with different operating thresholds of the Gerber. The more the Gerber works continuously, the higher the correlation between the two series. The correlation coefficient reaches 0.83 when the PVM Gerber is working more than 70% of the time.

Table 2. Monthly data during 2013–2018 and the correlation coefficients after removing the months when PVM Gerber was working less than the different thresholds from 30% to 90%.

PVM Gerber Working More than	Number of Monthly Points	Correlation Coefficient
0% (The whole serie)	72	0.46
30%	59	0.53
40%	57	0.55
50%	53	0.65
60%	49	0.73
70%	41	0.82
80%	37	0.89
90%	31	0.92

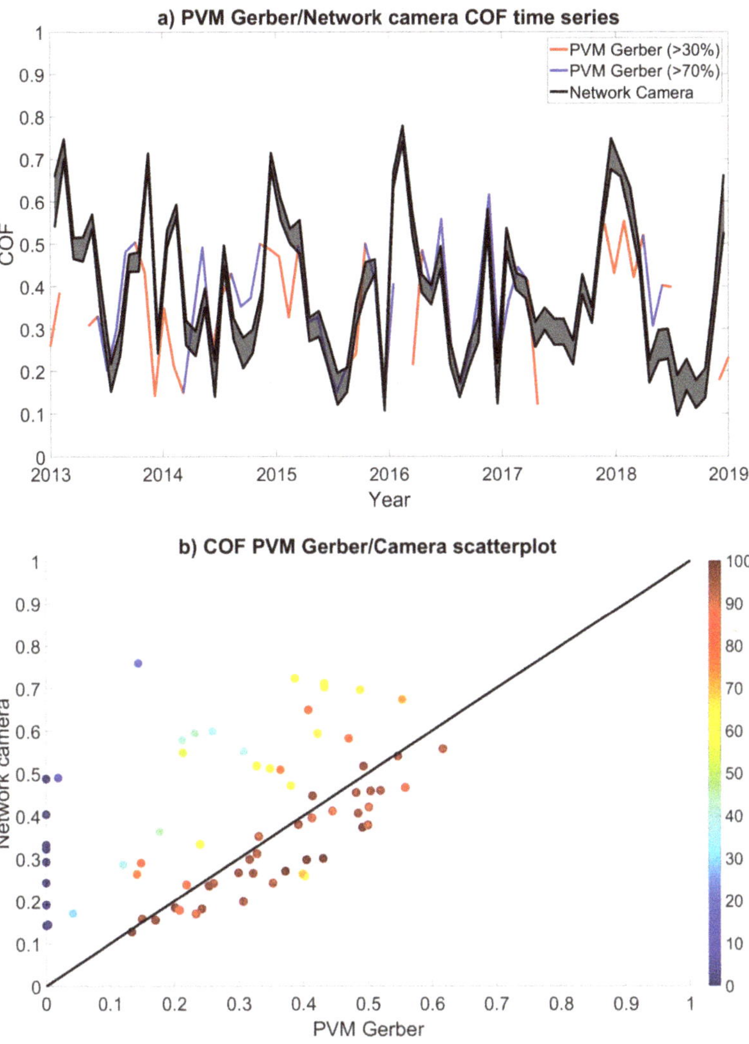

Figure 4. (**a**) Cloud occurrence frequency (COF) monthly time series calculated from camera images in black, with PVM Gerber in red. The lower limit of the camera curve is obtained with the threshold value 5.5, and the upper limit is obtained by adding the cases classified as "uncertain" with threshold values between 5.5 and 6.5. The PVM Gerber points with less than 30% of validated available data have been removed, and the PVM Gerber points with more than 30% and 70% are in red and blue, respectively. (**b**) A scatterplot of the monthly cloud occurrence from the camera images (threshold value at 6) as a function of PVM Gerber. The color is the percentage of validated PVM Gerber data per month, and the line y = x is in black.

5. Diurnal, Seasonal, and Inter-Annual Variations of the COF at PUY

The camera data processing methodology for the estimation of the COF presented in Section 3 and validated in Section 4 was applied to all the PUY images from 2013 to 2018 to analyze the temporal variations at different time scales: diurnal, seasonal, and inter-annual. Figure 5 presents the mean COF daily variation separated by seasons and years.

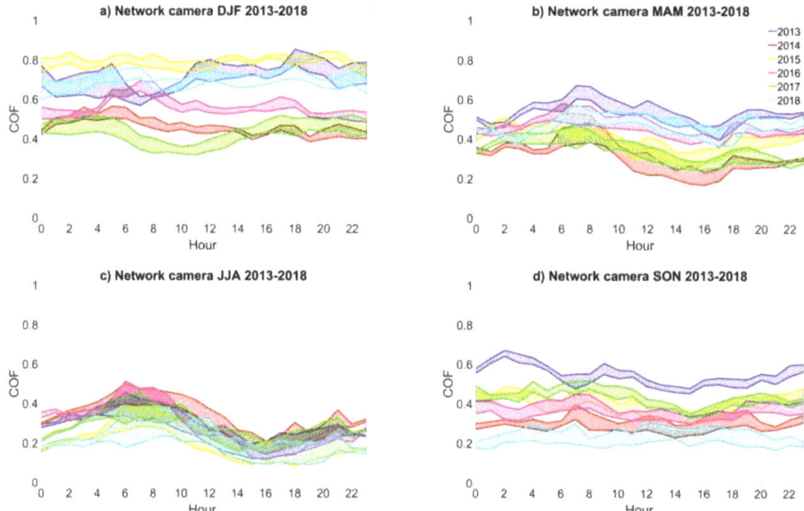

Figure 5. (**a**) Network camera DJF 2013-2018; (**b**) Network camera MAM 2013-2018; (**c**) Network camera JJA 2013-2018; (**d**) Network camera SON 2013-2018. COF obtained from the camera image processing separated by seasons and years as a function of the hours of the day (UTC). For each year and season, the uncertainty interval has been estimated similarly to Figure 4a. The correspondence between the colors and years is given in the top-right corner.

The main climatological features we can address are the following:

- There is a clear diurnal cycle during summer, with a larger cloud occurrence from 0 to 11 UT (mean value = 0.30) than from 12 to 23UT (mean value = 0.18), a less pronounced diurnal cycle during spring, and no diurnal cycle observed during winter and autumn. This feature could be linked to the seasonal changes in the day/night contrast of the boundary layer height (BLH), with a higher BLH observed during daytime in summer and spring [32]. During summer and spring, more daytime clouds may occur above the PUY station and are thus not detected by the camera.
- A strong seasonal cycle is observed, with the largest values in winter (mean value = 0.60), followed by spring (mean value = 0.41), autumn (mean value = 0.37), and summer (mean value = 0.24).
- The inter-annual variability depends on the season. This value can be estimated by calculating the daily mean of the difference between the maximum value and the minimum value of cloud occurrence. We found the largest inter-annual variability in winter (0.38), then autumn (0.32) and spring (0.23), with the lowest inter-annual variability in summer (0.14). The inter-annual variability does not depend on the hour of the day. The larger variability in winter could be simply due to the fact that there are more clouds in the winter. The low values of cloud cover in 2018 could be linked to the heat wave episode observed in Western Europe.

6. Comparisons with Larger Scale Cloud Parameters

It is interesting to compare the COF calculated from the camera image analysis (which corresponds to a local scale) to other parameters representative of the cloud conditions on a larger scale. In this section, we provide a preliminary comparison between the camera analysis results with different cloud cover parameters on a larger scale, in order to evaluate which product is closest to a local measure. This is also an estimation of how the PUY measurements are representative of larger scale cloud cover (i.e., how the measurements are biased by local orographic cloud formation).

6.1. CATS

Figure 6 shows the camera COF averaged over the CATS measurement periods (2015–2017), compared to the CF obtained from CATS.

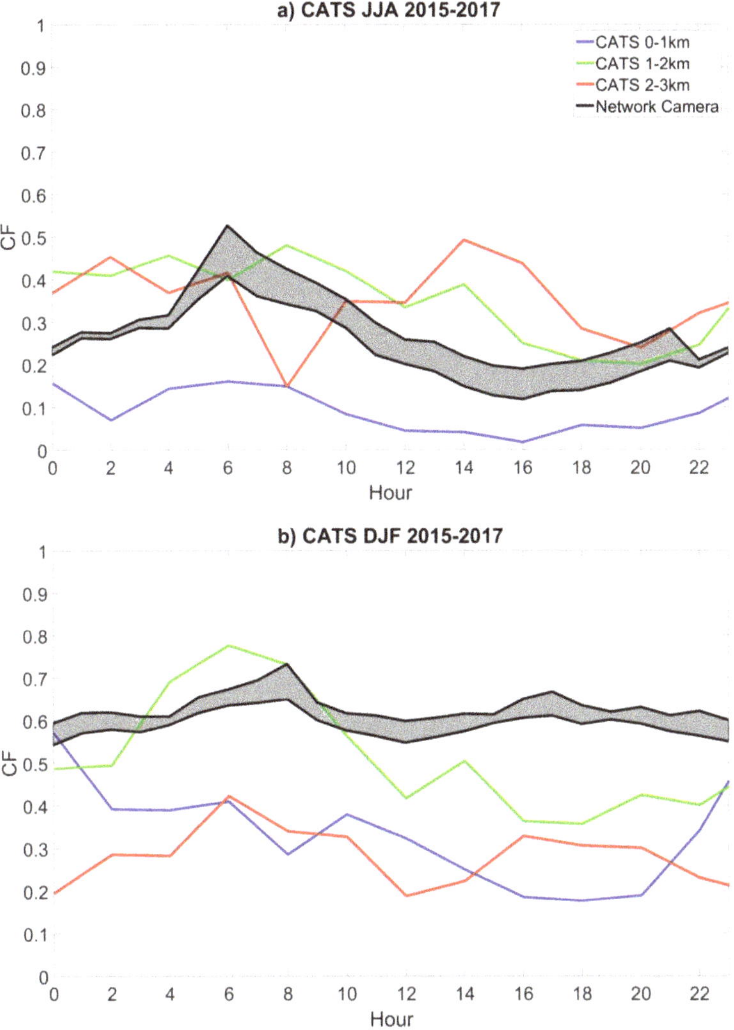

Figure 6. CF as function of time of the day (UT), extracted from the cloud-aerosol transport system (CATS) in summer 2015–2017 (**a**) and winter 2015–2017 (**b**) for three vertical ranges above sea level, 0–1 km in blue, 1–2 km in green, and 2–3 km in red, compared to the COF deduced from the PUY camera in black. The sea cloud camera situations have been added to the cloud situation for this comparison. The lower limit of the camera cloud occurrence (black curve) is obtained with a threshold value of 5.5, and the upper limit is obtained by adding the cases classified as "uncertain" (with a threshold value between 5.5 and 6.5).

We found that the closest CATS CF values to the camera COF are those of the 1–2 km layer (green curve), which is consistent with the altitude of the PUY roughly in the middle of this altitude range.

The results are presented by separating the database into two seasons, summer for the months of June to August (Figure 6a) and winter for the months of December to February (Figure 6b). The camera CF is of the same order of magnitude for both types of data, but CATS shows a larger diurnal variation, with CF values larger than the camera cloud occurrence values from 3 to 9 UT during winter and the opposite during the rest of the day. The camera does not show such daily variability during winter; the value of the CF is of the same order of magnitude during night and day (around 0.6). In summer, the daily variability of the CATS CF is less important than in winter, bringing the daily evolution closer to that of the camera. We found that the CF of CATS is, on average, 11% larger than that the camera cloud occurrence. In winter (Figure 6b), the CATS results show that the CF is largest at the PUY altitude (1–2 km). At higher and lower altitudes, the CF is significantly less. During summertime (Figure 6a), the CF at lower altitudes falls down to 0.1, while at higher altitudes (1–2 and 2–3 km), the CATS CF is quite similar to the camera COF.

6.2. Meteosat-SEVIRI

We compared the COF values determined by the camera image analysis with those extracted from METEOSAT SEVIRI on the nearest pixel to PUY and also on a larger area of around 200 km for different vertical levels of the cloud top altitudes (Figure 7).

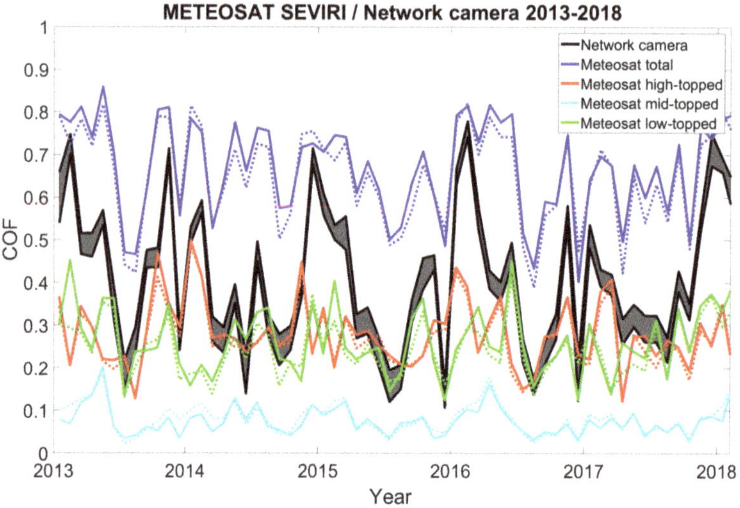

Figure 7. Total, high-topped, mid-topped, and low-topped COF (in blue, red, cyan, and green, respectively), extracted from METEOSAT-SEVIRI on the nearest pixel over PUY (continuous lines), as well as on a larger region between 44.6 and 46.9° N and between 2 and 4° E (dashed lines), compared to the monthly mean of the camera COF in black.

The SEVIRI total COF is always larger than the COF deduced from the camera analysis. The difference is 0.25 in average, which probably corresponds to the cases when the cloud base is higher than the altitude of PUY or when the cloud top is lower than the PUY altitude.

The seasonality of COF obtained by the camera analysis (maximum clouds in winter and minimum in summer) is also seen in the METEOSAT-SEVIRI total COF but with a smaller amplitude. For example, the difference between winter and summer on the total COF is only between 0.10 and 0.30. The METEOSAT-SEVIRI low-topped cloud curve is the closest to the camera COF curve (correlation coefficient of 0.60), but the winter maximums are underestimated. In winter, peaks in the webcam COF curve coincide with peaks in the low-topped COF serie but also with peaks in the high-topped COF serie. This is in agreement with the more frequent occurrence of large cloud systems associated with disturbances during this season [28]. In this large-scale cloud system, thick upper clouds prevent the detection of low clouds with METEOSAT-SEVIRI.

Otherwise, for all altitude levels, the METEOSAT-SEVIRI curve corresponding to the nearest pixel over PUY is close to the curve corresponding to the 200 km domain. This suggests that the small-scale cloud cover around the top of PUY has less influence on the METEOSAT-SEVIRI cloud parameters than large scale cloud cover, regardless of the altitude range.

6.3. All-Sky Camera

We have also compared the COF deduced from the camera analysis to the CF calculated with the algorithm ELIFAN [18] applied to the all-sky camera images taken at the Cézeaux site from December 2015 to June 2019 (Figure 8).

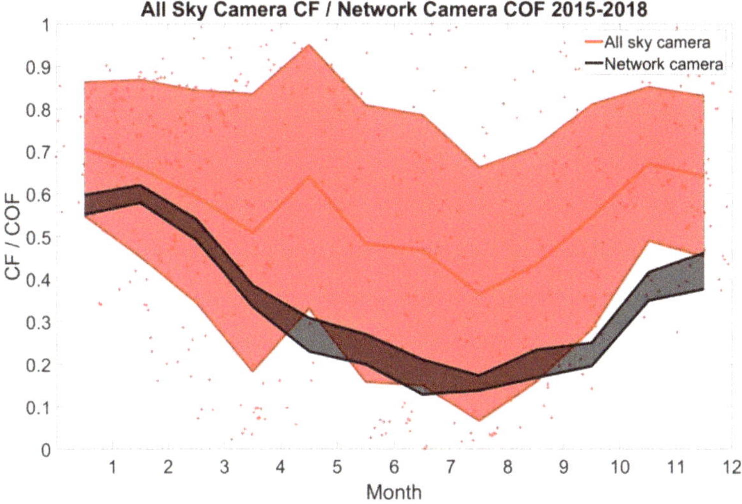

Figure 8. Daily (red crosses) and monthly +/− one sigma (red lines) means of the CF calculated with the algorithm ELIFAN applied to the all-sky camera images taken at the Cézeaux site from December 2015 to June 2019, compared to the monthly mean of the COF from the camera analysis in black.

The ELIFAN CF parameter presents larger variability compared to the local camera COF but follows a similar seasonal evolution, with larger mean values in winter (0.67) than in summer (0.44). The all-sky camera CF is determined locally but integrates the cloud cover over a larger area than the cloud occurrence at PUY. In agreement with the results presented in Section 6.2, the fact that the ELIFAN and camera parameters follow a similar evolution suggests that the COF at PUY is representative of large-scale processes.

6.4. ECMWF ERA-5

We have also compared the COF deduced from the camera analysis to the fraction of the cloud cover parameter of the ERA-5 dataset recently provided by ECMWF, interpolated on the closest point to PUY at a 850 hPa pressure level. Figure 9 shows the time series of the monthly averages of these two parameters. The two time series follow highly correlated variations (correlation coefficient 0.94) but with a mean shift of 19% (up to 40% for some winter months). This result suggests that the local cloud cover over a mountain site is probably underestimated by the ECMWF ERA-5 reanalysis. However, the monthly average plus one standard deviation is very close to the COF camera curve.

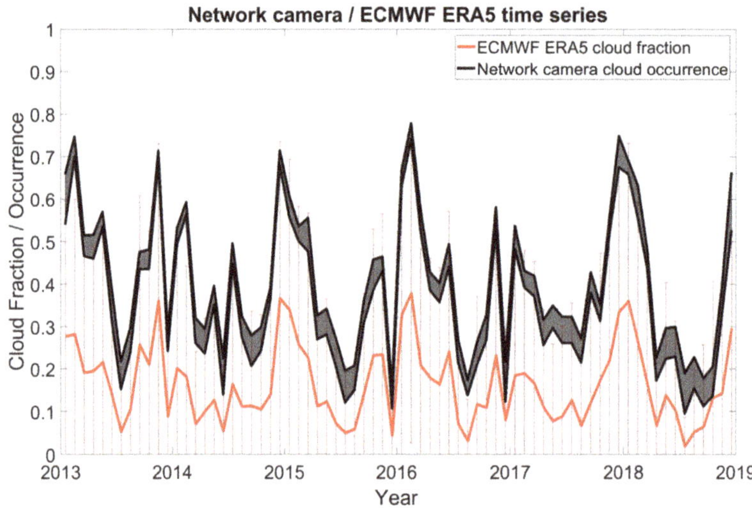

Figure 9. Time series of the monthly mean and standard deviation of the ERA 5 ECMWF fraction of the cloud cover interpolated on the point 45.77° N, 2.95° E on the 850 hPa pressure level in red, compared to the monthly mean of the camera COF in black.

7. Conclusions

The objective of this work was to show the feasibility of the detection of cloud occurrence on an altitude site from a simple, low-cost, and automatic algorithm based on an analysis of camera images. The comparison of the results of our algorithm with in-situ microphysical data (PVM Gerber) are in good agreement.

The application of this algorithm to the PUY camera over the period 2013–2018 shows that a cloud diurnal cycle is observable only in summer and spring (cloud occurrence is larger during the night than during the day), but this may be because high level daytime clouds during this season are not detected by the camera. There is a strong seasonal cycle with larger values of cloud occurrence in winter than in summer. The inter-annual variability is also high, although it is globally smaller than the seasonal cycle; however, it is also larger in winter than in summer. Specific meteorological conditions could probably explain this inter-annual variability. A more in-depth study of the relationship between the weather conditions and the inter-annual variability of the cloud occurrence frequency could be investigated in the future.

The seasonal variability of clouds was also found from larger scale cloud detection techniques: the all-sky imager located at Cézeaux, 10 km far from PUY, from space observations (CATS, Meteosat), and from global model outputs (ECMWF ERA-5). The one by one comparison of different techniques provides further information on the representativeness of the PUY site.

The comparison with CATS CF shows that, in winter, the CATS CF is of the same order of magnitude as the camera COF, with a larger daily variability for CATS, and, in summer, the daily variability of the CATS CF is less important than in winter, producing daily evolution closer to that of the camera COF. This observation confirms that daytime summertime clouds may develop above the site and are seen by the camera.

A comparison with METEOSAT-SEVIRI products shows that the low-topped COF is the closest parameter to the camera COF, but the winter maximum is underestimated. This is due at least in part to the fact that when high clouds are occurring, low clouds cannot be detected by METEOSAT-SEVIRI.

The comparison with the monthly fraction of cloud cover from the ECMWF ERA-5 reanalysis shows extremely correlated variations (a coefficient correlation of 0.93) but a mean shift of 19%, with larger values for the camera COF than the ECMWF ERA-5 fraction of cloud cover. Despite this shift, the camera COF values remain in the range of ERA-5 +/− one standard deviation.

The CF estimated with the all-sky image at Cézeaux presents a larger variability than the camera COF but follows a similar seasonal evolution, with larger values in winter (0.67) than in summer (0.43). In agreement with the other large scale CF parameters, this suggests that the cloud occurrence PUY is mainly governed by large scale processes.

This automatic determination of cloud occurrence is of interest in characterizing observation sites, especially those that are not yet equipped with instruments for the microphysical characterization of clouds. This methodology can also be applied to other high-altitude sites, such as the Maïdo station (Réunion Island, Southern Indian Ocean [33]), where three cameras oriented toward the west, south-west, and east are in operation (Figure 10). The application to camera images from other stations would probably need an adaptation of the choice of windows and variability detection thresholds, which depend on the angle of viewing and on the quality of the images.

Figure 10. Examples of images of a camera in operation at Maïdo station (Réunion Island, 2160 m above the sea level, towards the south-east—clear sky and cloudy (top), and towards the south west during night and day (bottom), where the algorithm of detection of the CF could be applied in the future.

Another future plan could be to adapt Babari's method [34] to calculate atmospheric visibility coefficients from the altitude site cameras in order to classify the cloud occurrence related to the cameras' opacity.

Few altitude sites are equipped with complex and expensive cloud microphysics instruments, but many are equipped with simple cameras. The application of this method to a sufficient number of sites could be used to quantify the sub-grid variability of cloud cover as seen by most satellites or global models.

Author Contributions: Conceptualization, J.-L.B.; data curation, P.C., J.-M.P., N.M., V.N., and G.S.; formal analysis, J.-L.B., A.B., N.M., V.N., and G.S.; funding acquisition, J.-L.B., L.D., and N.M.; investigation, J.-L.B., A.B., K.S., N.M., V.N., and G.S.; methodology, J.-L.B.; project administration, J.-L.B.; resources, P.C., J.-M.P., N.M.; V.N., G.S.; F.G., and G.P.; software, J.-L.B. and A.B.; supervision, J.-L.B.; validation, J.-L.B.; visualization, J.-L.B.; writing—original draft, J.-L.B.; writing—review and editing, J.-L.B., K.S., L.D., N.M.,V.N., G.S., and V.D.

Funding: This research received no external funding.

Acknowledgments: This study benefited from discussions during the workshops ACTRIS-FR and EECLAT and the seminar given to the laboratory LaMP. Régis Dupuy and Pierre Coutris are acknowledged for suggesting useful references, Sandra Banson for extracting ECMWF ERA 5 reanalysis, and Nicolas Pascal for reprocessing ELIFAN calculations.

Conflicts of Interest: The authors declare no conflict of interest.

References

1. Hartmann, D.L.; Doelling, D. On the net radiative effectiveness of clouds. *J. Geophys. Res. Atmos.* **1991**, *96*, 869–891. [CrossRef]

2. Van Baelen, J.; Reverdy, M.; Tridon, F.; Labbouz, L.; Dick, G.; Bender, M.; Hagen, M. On the relationship between water vapour field evolution and the life cycle of precipitation systems. *Q. J. R. Meteorol. Soc.* **2011**, *137*, 204–223. [CrossRef]

3. Bony, S.; Colman, R.; Kattsov, V.M.; Allan, R.P.; Bretherton, C.S.; Dufresne, J.-L.; Hall, A.; Hallegatte, S.; Holland, M.M.; Ingram, W.; et al. How well do we understand and evaluate climate change feedback processes? *J. Clim.* **2006**, *19*, 3445–3482. [CrossRef]

4. Ceppi, P.; Brient, F.; Zelinka, M.D.; Hartmann, D.L. Cloud feedback mechanisms and their representation in global climate models. *Wiley Interdiscip. Rev. Clim. Chang.* **2017**, *8*, e465. [CrossRef]

5. Wallace, J.M.; Hobbs, P.V. 6—Cloud Microphysics. In *Atmospheric Science*, 2nd ed.; Wallace, J.M., Hobbs, P.V., Eds.; Academic Press: San Diego, CA, USA, 2006; pp. 209–269. ISBN 978-0-12-732951-2.

6. Renard, P.; Canet, I.; Sancelme, M.; Wirgot, N.; Deguillaume, L.; Delort, A.-M. Screening of cloud microorganisms isolated at the Puy de Dôme (France) station for the production of biosurfactants. *Atmos. Chem. Phys.* **2016**, *16*, 12347–12358. [CrossRef]

7. Köhler, H. The nucleus in and the growth of hygroscopic droplets. *Trans. Faraday Soc.* **1936**, *32*, 1152–1161. [CrossRef]

8. ACTRIS Objectives Webpage. Available online: https://www.actris.eu/About/ACTRIS/Objectives.aspx (accessed on 12 December 2019).

9. Guyot, G.; Gourbeyre, C.; Febvre, G.; Shcherbakov, V.; Burnet, F.; Dupont, J.-C.; Sellegri, K.; Jourdan, O. Quantitative evaluation of seven optical sensors for cloud microphysical measurements at the Puy-de-Dôme Observatory, France. *Atmos. Meas. Tech.* **2015**, *8*, 4347–4367. [CrossRef]

10. Deguillaume, L.; Charbouillot, T.; Joly, M.; Vaïtilingom, M.; Parazols, M.; Marinoni, A.; Amato, P.; Delort, A.-M.; Vinatier, V.; Flossmann, A.; et al. Classification of clouds sampled at the puy de Dôme (France) based on 10 yr of monitoring of their physicochemical properties. *Atmos. Chem. Phys.* **2014**, *14*, 1485–1506. [CrossRef]

11. Vaïtilingom, M.; Attard, E.; Gaiani, N.; Sancelme, M.; Deguillaume, L.; Flossmann, A.I.; Amato, P.; Delort, A.-M. Long-term features of cloud microbiology at the puy de Dôme (France). *Atmos. Environ.* **2012**, *56*, 88–100. [CrossRef]

12. Li, J.; Wang, X.; Chen, J.; Zhu, C.; Li, W.; Li, C.; Liu, L.; Xu, C.; Wen, L.; Xue, L.; et al. Chemical composition and droplet size distribution of cloud at the summit of Mount Tai, China. *Atmos. Chem. Phys.* **2017**, *17*, 9885–9896. [CrossRef]

13. Van Pinxteren, D.; Fomba, K.W.; Mertes, S.; Müller, K.; Spindler, G.; Schneider, J.; Lee, T.; Collett, J.L.; Herrmann, H. Cloud water composition during HCCT-2010: Scavenging efficiencies, solute concentrations, and droplet size dependence of inorganic ions and dissolved organic carbon. *Atmos. Chem. Phys.* **2016**, *16*, 3185–3205. [CrossRef]

14. Bianco, A.; Deguillaume, L.; Vaïtilingom, M.; Nicol, E.; Baray, J.-L.; Chaumerliac, N.; Bridoux, M. Molecular Characterization of Cloud Water Samples Collected at the Puy de Dôme (France) by Fourier Transform Ion Cyclotron Resonance Mass Spectrometry. *Environ. Sci. Technol.* **2018**, *52*, 10275–10285. [CrossRef] [PubMed]

15. Baray, J.-L.; Deguillaume, L.; Colomb, A.; Sellegri, K.; Freney, E.; Rose, C.; Van Baelen, J.; Pichon, J.-M.; Picard, D.; Fréville, P.; et al. Cézeaux-Aulnat-Opme-Puy De Dôme: A multi-site for the long term survey of the tropospheric composition and climate change. *Atmos. Meas. Tech. Discuss.* **2019**. [CrossRef]
16. Axis Webpage. Available online: https://www.axis.com/en (accessed on 12 December 2019).
17. OPGC Webcam Page. Available online: http://wwwobs.univ-bpclermont.fr/opgc/webcam.php (accessed on 12 December 2019).
18. Lothon, M.; Barnéoud, P.; Gabella, O.; Lohou, F.; Derrien, S.; Rondi, S.; Chiriaco, M.; Bastin, S.; Dupont, J.-C.; Haeffelin, M.; et al. ELIFAN, an algorithm for the estimation of cloud cover from sky imagers. *Atmos. Meas. Tech.* **2019**, *12*, 5519–5534. [CrossRef]
19. Gerber, H. Liquid water content of fogs and hazes from visible light scattering. *J. Clim. Appl. Meteorol.* **1984**, *23*, 1247–1252. [CrossRef]
20. Gerber, H. Direct measurement of suspended particulate volume concentration and far-infrared extinction coefficient with a laser-diffraction instrument. *Appl. Opt.* **1991**, *30*, 4824–4831. [CrossRef]
21. McGill, M.J.; Yorks, J.E.; Scott, V.S.; Kupchock, A.W.; Selmer, P.A. The Cloud-Aerosol Transport System (CATS): A technology demonstration on the International Space Station. *Lidar Remote Sens. Environ. Monit.* **2015**, *96*, 1–6.
22. Pauly, R.M.; Yorks, J.E.; Hlavka, D.L.; McGill, M.J.; Amiridis, V.; Palm, S.P.; Rodier, S.D.; Vaughan, M.A.; Selmer, P.A.; Kupchock, A.W.; et al. Cloud Aerosol Transport System (CATS) 1064 nm calibration and validation. *Atmos. Meas. Tech. Discuss.* **2019**, *12*, 6241–6258. [CrossRef]
23. Noel, V.; Chepfer, H.; Chiriaco, M.; Yorks, J. The diurnal cycle of cloud profiles over land and ocean between 51° S and 51° N, seen by the CATS spaceborne lidar from the International Space Station. *Atmos. Chem. Phys.* **2018**, *18*, 9457–9473. [CrossRef]
24. Data Product Catalog Release 3.0. Available online: https://cats.gsfc.nasa.gov/media/docs/CATS_Data_Products_Catalog.pdf (accessed on 12 December 2019).
25. EUMETSAT NWC SAF Homepage. Available online: http://www.nwcsaf.org/ (accessed on 12 December 2019).
26. Derrien, M.; Le Gléau, H. MSG/SEVIRI cloud mask and type from SAFNWC. *Int. J. Remote Sens.* **2005**, *26*, 4707–4732. [CrossRef]
27. Derrien, M.; Le Gléau, H. Improvement of cloud detection near sunrise and sunset by temporal-differencing and region-growing techniques with real-time SEVIRI. *Int. J. Remote Sens.* **2010**, *31*, 1765–1780. [CrossRef]
28. Sèze, G.; Pelon, J.; Derrien, M.; Le Gléau, H.; Six, B. Evaluation against CALIPSO lidar observations of the multi-geostationary cloud cover and type dataset assembled in the framework of the Megha-Tropiques mission. *Q. J. R. Meteorol. Soc.* **2015**, *141*, 774–797. [CrossRef]
29. Albergel, C.; Dutra, E.; Munier, S.; Calvet, J.-C.; Munoz-Sabater, J.; de Rosnay, P.; Balsamo, G. ERA-5 and ERA-Interim driven ISBA land surface model simulations: Which one performs better? *Hydrol. Earth Syst. Sci.* **2018**, *22*, 3515–3532. [CrossRef]
30. Matlab Mathworks Homepage. Available online: https://www.mathworks.com/products/matlab.html (accessed on 12 December 2019).
31. Python Homepage. Available online: https://www.python.org (accessed on 12 December 2019).
32. Farah, A.; Freney, E.; Chauvigné, A.; Baray, J.-L.; Rose, C.; Picard, D.; Colomb, A.; Hadad, D.; Abboud, M.; Farah, W.; et al. Seasonal variation of aerosol size distribution data at the puy de Dôme station with emphasis on the boundary layer/free troposphere segregation. *Atmosphere* **2018**, *9*, 244. [CrossRef]
33. Baray, J.-L.; Courcoux, Y.; Keckhut, P.; Portafaix, T.; Tulet, P.; Cammas, J.-P.; Hauchecorne, A.; Godin Beekmann, S.; De Mazière, M.; Hermans, C.; et al. Maïdo observatory: A new high-altitude station facility at Reunion Island (21° S, 55° E) for long-term atmospheric remote sensing and in situ measurements. *Atmos. Meas. Tech.* **2013**, *6*, 2865–2877. [CrossRef]
34. Babari, R.; Hautière, N.; Dumont, É.; Paparoditis, N.; Misener, J. Visibility monitoring using conventional roadside cameras—Emerging applications. *Transp. Res. Part C Emerg. Technol.* **2012**, *22*, 17–28. [CrossRef]

Article

An Exceptional Case of Freezing Rain in Bucharest (Romania)

Simona Andrei [1,*], Bogdan Antonescu [1], Mihai Boldeanu [1,2], Luminiţa Mărmureanu [1], Cristina Antonia Marin [1,3], Jeni Vasilescu [1] and Dragoş Ene [1]

[1] National Institute of Research and Development for Optoelectronics INOE2000, Str. Atomiştilor 409, Măgurele, RO077125 Ilfov, Romania; bogdan.antonescu@inoe.ro (B.A.); mihai.boldeanu@inoe.ro (M.B.); mluminita@inoe.ro (L.M.); cristina.marin@inoe.ro (C.A.M.); jeni@inoe.ro (J.V.); dragos.ene@inoe.ro (D.E.)

[2] Faculty of Electronics, Telecommunications and Information Technology, Politehnica University of Bucharest, Spl. Independentei 313, RO061071 Bucharest, Romania

[3] Department of Physics, Politehnica University of Bucharest, Str. Spl. Independentei 313, RO060042 Bucharest, Romania

[*] Correspondence: simona.andrei@inoe.ro

Received: 4 October 2019; Accepted: 30 October 2019; Published: 1 November 2019

Abstract: A high-impact freezing rain event affected parts of southeastern Romania on 24–26 January 2019. The freezing rain caused extensive damages in Bucharest, the capital city of Romania. The meteorological analysis highlighted the presence of a particular synoptic pattern involving a high-pressure system advecting cold air mass at low levels, while at mid-levels a warm and humid intrusion was associated with a low-pressure system of Mediterranean origin. At Bucharest, the vertical profiles from ERA5 and radiosondes emphasized the presence of a thick warm layer between 1000–1400 m above the re-freezing layer close to the surface. A climatology of freezing rain events in Bucharest was built to understand the frequency and intensity of this phenomenon. On average, there were approximately 5 observations of freezing rain in Bucharest per year between 1980–2018. The number of consecutive freezing rain days was used as a proxy for the event severity. Moderate-duration events (2 consecutive days) represented 16 periods of all 59 non-overlapping freezing rain periods in Bucharest and long-duration events (3 consecutive days) represented 3 periods. The monthly distribution showed that freezing rain occurs more frequently between December–February with a maximum in December. The moderate and long-duration freezing rain events were associated with two main sub-synoptic patterns related to the Carpathians lee cyclogenesis.

Keywords: freezing rain; high-impact meteorological event; Carpathian lee cyclogenesis; climatology

1. Introduction

Freezing rain—"rain that falls in liquid form but freezes upon impact to form a coating of glaze upon the ground and on exposed objects" [1]—is a meteorological phenomena more frequently observed in Europe during the winter months (i.e., December–February) [2,3]. The extensive ice accumulation on the ground associated with freezing rain can lead to a significant impact on human activities, surface and air transportation, infrastructure and electrical power transmission [4–7]. Also, freezing rain events associated with a small accumulation of ice or those that occur over a short period can also result in transportation difficulties and an increase in the number of accidents [8].

Previous studies have indicated that freezing rain occurs when an elevated warm layer (i.e., >0 °C) is located above a cold layer (i.e., <0 °C) close to the surface [4,9–18]. If the warm layer is thick enough, the ice hydrometeors (i.e., snow) that fall through the layer can melt completely. Subsequently, as newly form liquid hydrometeors cross a subfreezing layer they become supercooled through conduction and evaporation. This occurs in environments in which the number of active ice nuclei is very low. Freezing rain can also occur even if a warm layer is not present [19–22]. For example, in a study based on 10 years of observations from weather stations in the continental United States, 25% of all freezing rain cases occurred without a warm layer [20]. For those cases, freezing precipitations were associated with the upward motion in a shallow saturated layer situated close to the surface. The temperature within the layer is too warm (i.e., ≥ -10 °C) for the ice nuclei to be active and the droplets grow by collision and coalescence. Thus, if the warm layer is present, then freezing rain is forming (i.e., diameter of the droplets >0.5 mm). If the warm layer is not present, then freezing drizzle is forming (i.e., diameter of the droplets <5 mm) [23].

Previous studies have analyzed the synoptic-scale patterns and local influences on freezing rain [24–26]. For example, surface observations, synoptic-scale weather maps, and radiosonde data were used to develop climatologies for freezing rain at six sites across the United States [27]. The results showed that freezing rain events were primarily driven by warm advection and their frequency was influenced by topographic features (i.e., height <1 km), water bodies (from small bays to oceans) and the dominant storm tracks during winter. Other studies based sounding associated with freezing rain events showed that (1) the depth of the elevated warm layer was between 500–2800 m, (2) the temperature in the warm layer varied between 1°C–10°C, (3) the depth of the cold layer was between 100–1400 m and (4) the temperature in the cold layer varied between -7 °C––1 °C [27]. Also for the United States, an analysis of hourly surface observations between 1976–1990 showed that freezing rain occurred on the cold side of stationary fronts (57% of all cases) or was associated with close surface cyclones (43% of all cases) [28]. The majority of the freezing rain events associated with cyclones (77%) occurred when the cyclone was east, southeast or south of the area with freezing rain reports. For Canada, 46 long-duration (i.e., ≥ 6 h) freezing rain events that occurred in Montreal between 1979–2008 were manually classified based on the 500-hPa trough axis in the west, central and east events [23]. Despite being driven by different synoptic scale features (e.g., strong surface anticyclone, quasi-stationary area of geostrophic frontogenesis) all events were characterized by cold northeasterly wind channelled in the St. Lawrence River valley. The strong low-level temperature inversions in the valley were associated with northeasterly surface winds.

In recent years, several impactful freezing rain events occurred in Europe. For example, a freezing precipitation event that occurred between 25–26 December 2010 in Moscow, Russian Federation affected the public (i.e., power lines were destroyed) and air transport (i.e., flights were cancelled at the Domodedovo Airport) [3]. In 2014, the freezing rain events that occurred between 31 January–4 February over the Balkan Peninsula resulted in 10 cm accumulation of ice in Slovenia and damages estimated at EUR 400 million [29,30]. Despite these major events, few studies were devoted to the climatology and synoptic-scale and local influences on freezing precipitation events. For example, freezing rain events (including freezing drizzle and ice pellets) from Europe between 1995–1998 were analyzed using surface observations [2]. The results showed that freezing precipitation is not an uncommon phenomenon. This type of event occurs more frequently in December–February and in regions characterized by a continental climate (i.e., Central Europe). Also, freezing rain and freezing drizzle are more frequently reported than ice pellets. Recently, a method for estimating the occurrence of freezing rain events based on gridded atmospheric data (i.e., ERA-Interim reanalysis) was developed [3]. The method, which was calibrated with SYNOP weather station observations, was used to develop a pan-European climatology of freezing rain for the period 1979–2014. The results showed that freezing rain events occur more frequently over central and Eastern Europe compared with northern Europe and coastal regions. Over central and Eastern Europe the freezing rain season is October–April with 1–2 events per year.

Despite freezing rain being a high impact hazardous event, very few studies have analyzed this type of weather event in Romania. For example, a study of the spatial and the temporal distribution of freezing rain events in Romania based on the surface observations between 1980–1999 concluded that the conditions associated with the occurrence of freezing rain "are quite difficult to fulfil and mostly depend on local factors" [31]. Recently, a high impact freezing rain event occurred between 24–26 January 2019 in Bucharest, the capital city of Romania and one of the most densely populated cities in Europe (i.e., 9265 inhabitant per km^2 [32]). Given the duration and impact of this event, the aim of this article is (1) to examine the event from a climatological perspective to better understand the frequency and severity of freezing rain in an urban area and (2) to determine the prevalent mesoscale processes associated with long duration freezing-rain events.

This article is organized as follows. The data sets and methods used for analyzing this particular event and its climatology are described in Section 2. Section 3.1 presents the effect of the freezing rain event in the largest city of Romania and its surroundings. The analysis of the phenomenon from a synoptic and mesoscale point of view is detailed in Section 3.2. To assess the impact of the event compared to previous ones, a climatology of freezing rain was performed together with a classification of synoptic patterns leading to freezing rain events in Bucharest (Section 4). Section 5 summarizes the results of this article.

2. Data

2.1. Renalysis

The meteorological analysis of the 24–26 January 2019 freezing rain event was performed using ECMWF (European Centre for Medium-Range Weather Forecasts) ERA5 hourly surface (i.e., mean sea level pressure, 2 m temperature) and upper air (i.e., geopotential height, temperature and wind at different pressure levels) reanalysis data with $0.25° \times 0.25°$ horizontal resolution ([33] (data available from https://cds.climate.copernicus.eu/#!/home, retrieved on 1 May 2019). The synoptic analysis was performed over a European domain situated between 35° N–55° N and 5° E–35° E longitude (Figure 1). To emphasize the influence of regional topography and the physical processes that favoured the freezing rain formation, a meso-α domain was defined between 16° E–34° E and 40° N–50° N (dark blue rectangle in Figure 1). For the local analysis, a meso-β domain was defined between 25.75° E–26.50° E and 44.25° N–44.75° N around Bucharest (black rectangle in Figure 1).

Local observations from the Romanian Atmospheric 3D Observatory (RADO, http://environment. inoe.ro/, accessed on 3 October 2019) located southwest of Bucharest, in Măgurele—a peri-urban area (Figure 1), were collected during the freezing rain event using a vertical pointed Micro Rain Radar (MRR, manufactured by Metek GmbH). The MRR is a Frequency Modulated Continuous Wave (FMCW) radar operating at 24 GHz, used to study, for example, liquid and solid precipitations [34,35] and bright band [36]. These observations were supplemented by radiosonde measurements extracted from the sounding database maintained by the University of Wyoming (http://weather.uwyo.edu/upperair/ sounding.html, accessed on 1 September 2019). During the event (00 UTC 24 January–00 UTC 27 January 2019), 7 soundings were available, twice a day at 00 UTC and 12 UTC (local time + 2 h).

2.2. Observations

To understand the climatological context of the freezing rain event, surface observation (SYNOP observations) from Global Surface Archives (World Meteorological Organization FM-12 SYNOP coded data [37]) between 1980–2018 were used. Data were extracted for three weather stations located in Bucharest (i.e., Băneasa, Filaret and Afumați). When analyzing the freezing rain reports from weather stations, some issues need to be considered. For example, some stations have reports for limited periods or the reports are missing overnight [22,38]. To address these issues, recent climatological studies of freezing rain in the United States were based only on data from stations that had reports for at least 80% of the annual hours for at least 10 out the 15 years studied [38] or for at least 30 out of 38 years studied [22]. When these criteria are applied to the three stations from Bucharest only Băneasa is identified as suitable for analysis. For Băneasa station, the reports were available every 3-h between 1980–1981 and hourly between 2007–2018. The remaining years (i.e., 1982–2006) had a combination of one hour and 3-hour reports. Between 1980–2018, Băneasa station had reports for more than 90% of the annual hours (considering hourly reports after 2007 and 3-h reports before 2007). In 2000, the Băneasa became a human-augmented automated station. Previous studies on the climatology of freezing rain in Europe did not include data from automated stations arguing that this type of station has a "reduced ability to distinguish [between] different types of precipitation, especially freezing rain and freezing drizzle" [3]. A more recent study indicated that human-augmented and non-augmented automated stations are consistent in their frequencies of freezing rain events [39]. Thus, in this article, we also include automated observations from Băneasa station.

For each SYNOP observation, the present weather consists of 100 codes describing the weather at the time of the observation or in the previous hour. Based on surface observation, the freezing rain events were identified as those observations containing the present weather code 24 (freezing rain within the previous hour), 66 (light freezing rain), and 67 (moderate to heavy freezing rain). During the study periods, 227,752 observations were available from Băneasa and of these, 183 were for freezing rain. To remove erroneous observations, we followed the methodology proposed by Kämäräinen et al. [3] and retained for analysis only the freezing rain observations with surface temperature between $-15\ ^\circ$C and $5\ ^\circ$C. Thus, the freezing rain dataset used in this article consists of 182 observations, the majority of the observations being for light freezing rain events (i.e., 80% of all freezing rain reports).

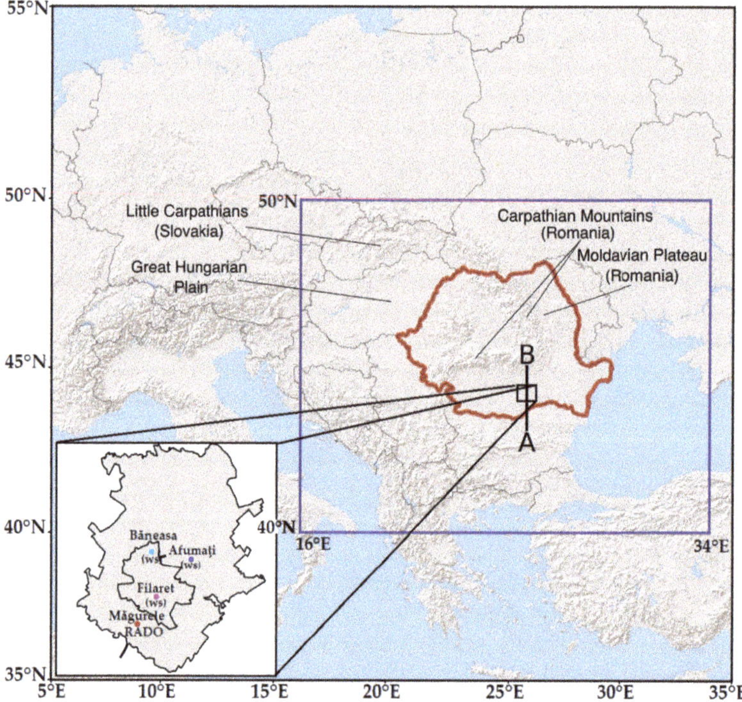

Figure 1. Study domain showing the synoptic and mesoscale (dark blue and black rectangles) areas over which analyses were performed. The red contour marks the Romanian territory. The black rectangle represents the location of Bucharest city and the measurement sites. The location of Romanian Atmospheric 3D Observatory (RADO)–Măgurele is indicated (red dot) together with the location of Băneasa (blue dot), Filaret (magenta dot) and Afumați (dark blue dot) weather stations. The A–B line shows the position of the cross-section in used in Section 3.2.

3. An Exceptional Freezing Rain Event in Bucharest

3.1. Description of the Event

On 24–26 January 2019 a freezing rain event affected parts of southeastern Romania including Bucharest, the capital city. During the event, the Romanian National Meteorological Administration issued 47 weather warnings for areas in southeastern Romania (Figure 1). The event resulted in the closing of several motorways connecting Bucharest with the rest of the country. At the *Henri Coandă* International Airport for three hours on 25 January the air traffic was closed. In Bucharest, the freezing rain affected 2622 trees (e.g., break off branches heavily coated with ice) and 759 cars were damaged (Figure 2). The public transport in Bucharest was interrupted due to power outages. According to the Romanian Emergency Service (personal communication) between 24–26 January, 154 calls were received for issues associated with freezing rain (i.e., 150 medical emergency and 4 car accidents).

Figure 2. (a) Freezing rain impact in Bucharest on 25 January 2019 and (b,c) ice accumulation in Măgurele (southwest of Bucharest). In (b) the ice layer on the branches was estimated at 2–4 cm.

3.2. The Synoptic and Mesoscale Analysis

On 24–26 of January 2019, Central and Eastern Europe were under the influence of an anticyclonic regime which determined the advection of polar air masses toward these regions. Southern Europe was directly influenced by an intense Mediterranean cyclonic activity (Figure 3a,c,e).

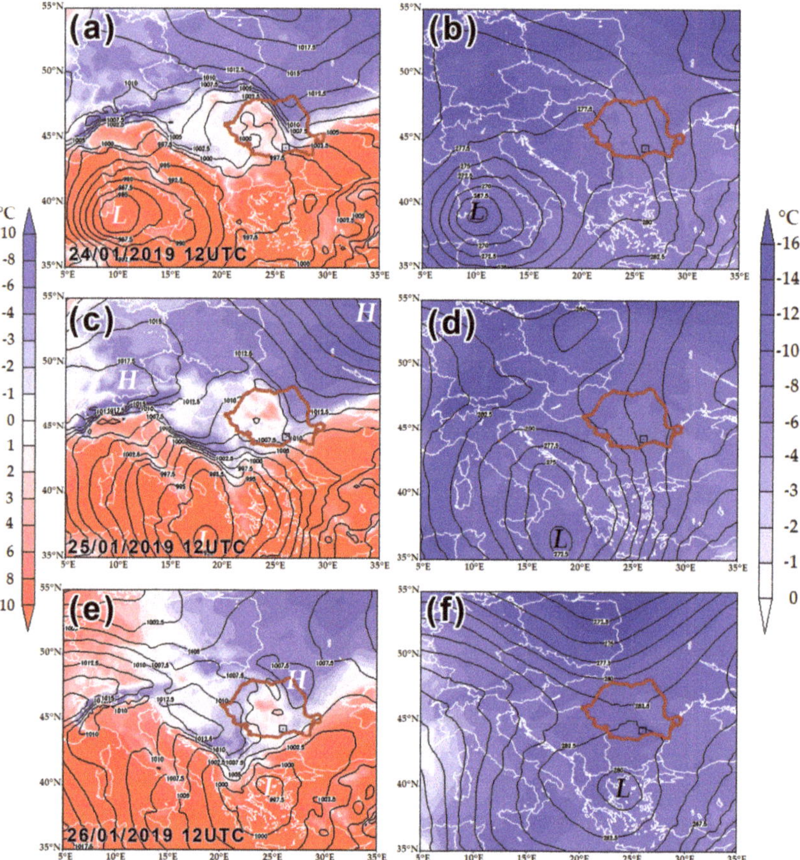

Figure 3. Synoptic analysis of freezing rain event between 24–26 January 2019 at 12 UTC, from ERA5. (**a,c,e**) shows mean sea level pressure (black contours) and 2 m temperature (shaded according to the scale). (**b,d,f**) shows the 700 hPa geopotential height (black contours) and temperature (shaded according the scale). The black rectangle marks the position of Bucharest.

The pressure systems evolution was influenced by the local topography. By their location and shape, the Carpathians Mountains played the role of a topographic barrier for the dense polar air, which was forced (at the lower levels) to follow the mountain curvature in its motion from north-east towards south-west. The high-pressure system formed two peri-Carpathians lobes, one over Eastern Romania (Moldavian Plateau) and another over Little Carpathians (western Slovakia) and Great Hungarian Plain [40]. Due to the cold air mass intrusion, the pre-existing warm air was lifted. This specific configuration triggers the Carpathians lee cyclogenesis [41]. Simultaneously, a large cut-off low, extended from the lower to upper levels, directed the frontal dynamics of the Mediterranean lows northward. This enhanced the advection of tropical humid air masses towards the Balkans (Figure 3b,d,f). As the Mediterranean lows approach the Balkans, the warm air was observed in a layer between 950–800 hPa, while close to the surface the polar air moved along the curvature of the Carpathians towards southern Romanian. The thermal analysis at meso-α scale reveals a contrast between the surface (1000 hPa) and the mid-troposphere (900 hPa) over southern Romania (Figure 4a–f).

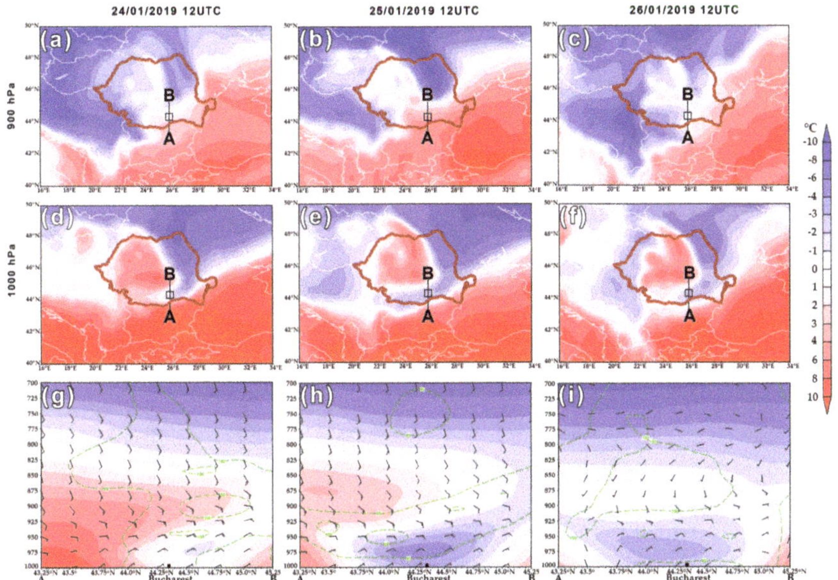

Figure 4. Regional and local analysis of freezing rain event during 24–26 January 2019 at 12 UTC from ERA5. (**a–c**) shows temperature at 900 hPa (°C, shaded according to the scale every 1 °C between −4 °C–4 °C and every 2 °C for the remaining contours between −10 °C–10 °C). (**d–f**) temperature at 1000 hPa (as in **a–c**). (**g–i**) the south-north vertical cross section at 26° E (A–B, Figure 1) showing temperature (as in a–c), relative humidity (green dotted contours, 5% increments) and wind profile (one pennant, full barb, and half-barb denote 25, 5 and 2.5 m s^{-1}, respectively).

The 26° E longitudinal vertical cross-section over Bucharest and surrounding areas emphasize the intrusion of cold air mass at the lower levels on a north-east wind component, while the warm and humid air mass advance at mid-levels having south-east and south-south-west wind component (Figure 4g–i) .

The time series of temperature profile over Bucharest area, performed for the closest ERA5 grid point to Măgurele city (Figure 5a) between 24 Jan 00 UTC and 27 Jan 00 UTC, emphasize the presence of a warm layer with a maximum temperature greater than 3 °C around 900 hPa. The cold layer (lower than −2 °C) is clearly shown between 24 Jan 12 UTC and 26 Jan 18 UTC (Figure 5a). This thermal structure leading to freezing rain shows similarity with the MRR measurements. The attenuated radar reflectivity shows the presence of a melting layer (i.e., bright band) between 1000–1400 m with values for radar reflectivity >20 dBz (Figure 5b).

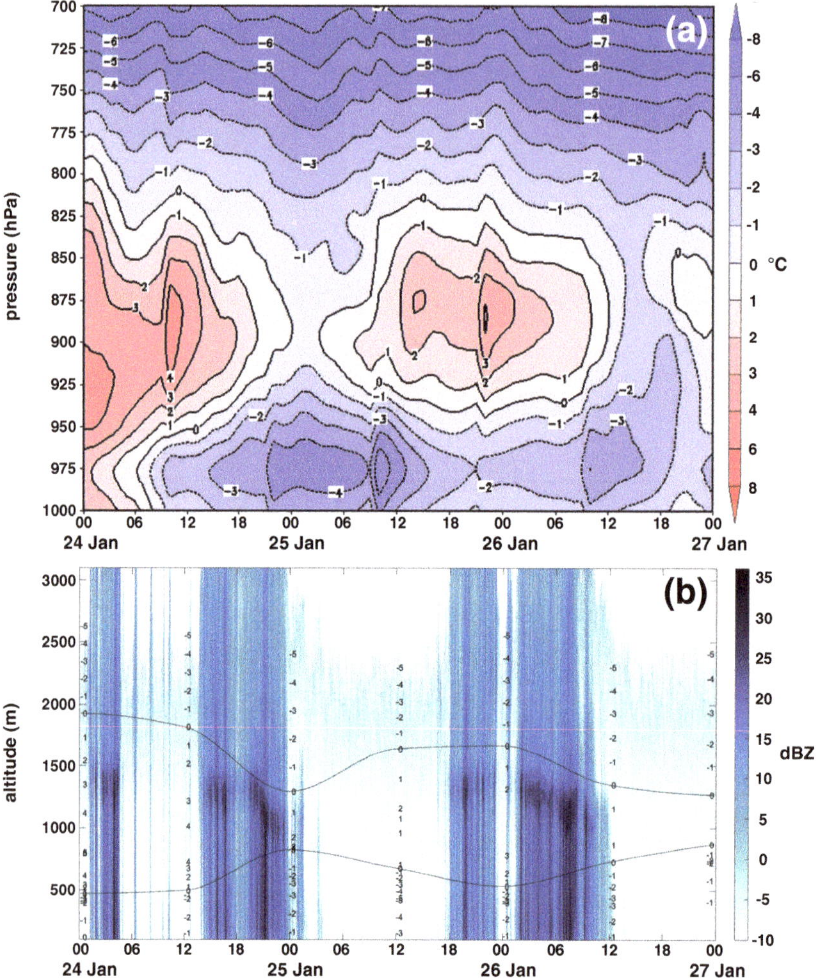

Figure 5. (**a**) Time series of temperature profiles between 24–26 January 2019 from ERA5 (shaded according to the scale, as in Figure 3a–c). (**b**) Micro Rain radar attenuated reflectivity (dBz) observations from Măgurele during the same period as in (**a**). Temperature profiles from the Bucharest sounding at 00 and 12 UTC are overlapped with the radar reflectively, the black contours separating positive and negative temperature values.

4. Climatology of Freezing Rain in Bucharest

On average, there were approximately 5 observations of freezing rain in Bucharest each year between 1980–2018 (Figure 6a). The number of freezing rain days (i.e., a day in which freezing rain was reported at least once) during the study period was 81 days corresponding to an average of 2 days per year (Figure 6a).

The monthly distribution of the freezing rain reports between 1980–2018 shows that this type of weather event occurs more frequently between December–February with a maximum in December (79 observations, representing 43% of all freezing rain observations, Figure 6b). A similar distribution was observed in the case of the freezing rain days with a maximum in December (31 freezing rain days, representing 38% of all freezing rain days).

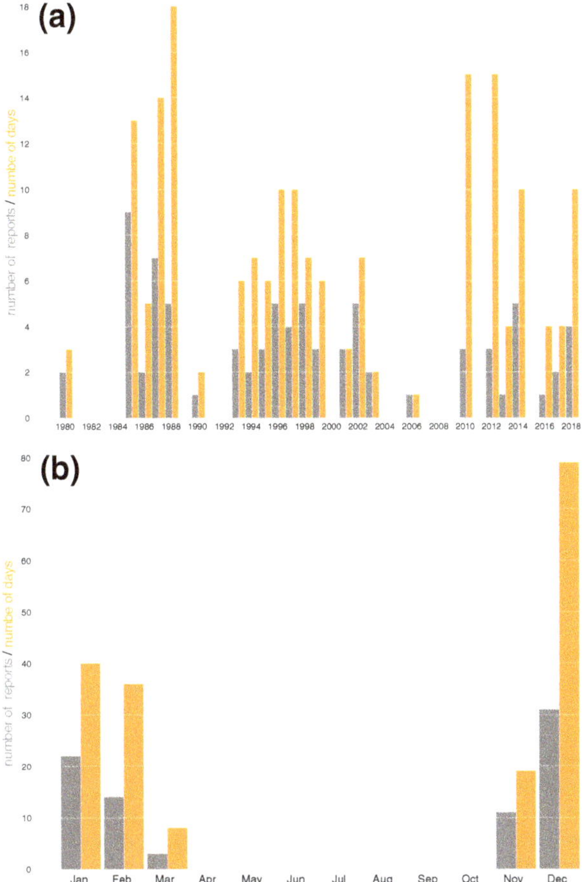

Figure 6. (**a**) The annual distribution of the number of freezing rain reports (orange bars) and the number of freezing rain days (i.e., days with a least one reports of freezing rain, grey bars) and (**b**) the monthly distribution of the number of freezing rain reports and the number of freezing rain days between 1980–2018 based on SYNOP observation from Băneasa station.

To account for the severity of freezing rain events in the United States and Canada, the duration of the event was considered as proxy [22]. Long-duration events being those in which at least six hours of freezing rain were reported. A similar approach could not be applied to the surface observations dataset used in this article, because our dataset contains hourly reports only for 2007–2018. The number of hours of freezing rain cannot be unambiguously determined from our dataset, given that between 1980–2016 the reports were made every three hours. Instead of using the hours of freezing rain, the number of consecutive freezing rain days was used as a proxy for the event severity. Thus, all freezing rain days between 1980–2018 in Bucharest were divided into 59 non-overlapping periods (Figure 7). Moderate-duration events (i.e., two consecutive freezing rain days) represented 27% (16 periods) of all non-overlapping periods and long-duration events (i.e., three or more consecutive freezing rain days) represented 5% (3 periods). Even if the thresholds (i.e., number of consecutive days) for moderate and long-duration events are subjective, we believe that they allow (1) the identification of high-impact freezing rain events (i.e., 2–3 consecutive days), and (2) the construction of a dataset with high-impact events. Thus, the event studied in this article had the largest number of hours with freezing rain, compared with other events that occurred between 2007–2018.

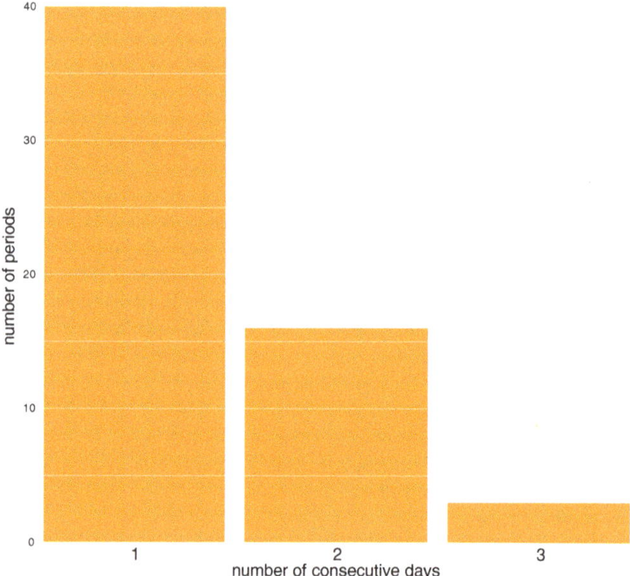

Figure 7. The distribution of non-overlaping periods with freezing rain between 1980–2018 based on SYNOP observation from Băneasa weather station.

The analyses of prolonged freezing rain events (i.e., moderate and long duration) recorded at Bucharest are driven by a sub-synoptic pattern related to the Carpathians lee cyclogenesis. The synoptic context which favours Carpathians lee cyclogenesis is associated with (1) a high-pressure system over Northern Romania originating either in Scandinavia or Eastern Europe and (2) a low-pressure system over Southern Romania with a Mediterranean origin. The high-pressure system moved southward simultaneously with the slow east-north-east transition of the Mediterranean low. Generally, the Carpathian lee cyclogenesis consists of a small size cyclone over Central and Southern Romania with a rapid evolution of 12–24 h. The processes are detectable only in the low troposphere (Figure 8c,d).

The 20 moderate and long-duration freezing rain events that occurred between 1980–2019 (including the case analyzed in this article) are associated with two main sub-synoptic patterns. These patters leading to Carpathian lee cyclogenesis are differentiated based on the origin of high-pressure system: Scandinavian origin (12 cases) (Figure 8a) and East European origin (7 cases) (Figure 8b) (Table 1). One event could not be included in the two main patters associated with freezing rain events over southern Romania, the *Lothar* windstorm from 26–27 December 1999 [42]).

The long-duration freezing rain events analyzed in this article were associated with the high pressure system of Scandinavian origin, that enables the fast southern advance of the arctic air-masses. This leads to a high-pressure gradient along the external part of Eastern Carpathian range (Figure 8a). The thermal contrast between the lower and mid-levels layer is higher compared with freezing rain cases associated with East European anticyclones (Figure 8b).

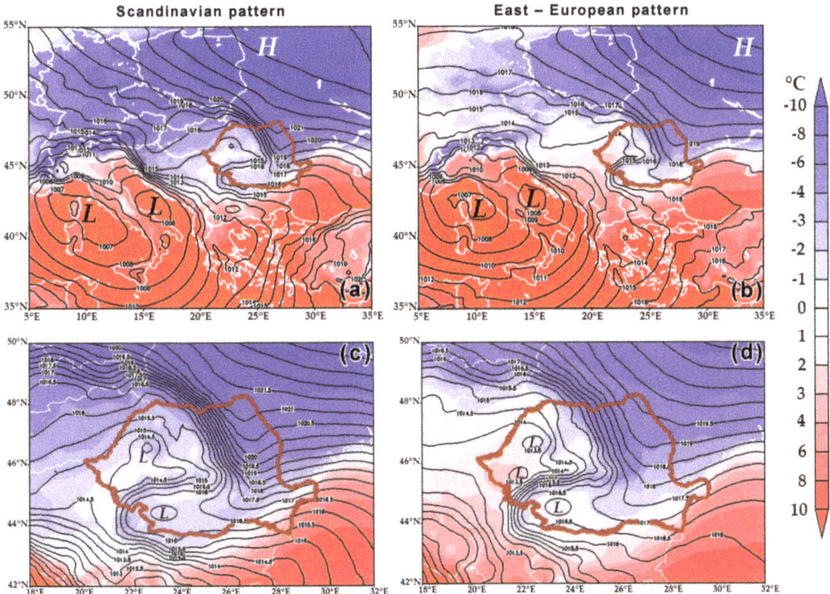

Figure 8. The composite maps of the two main synoptic patterns for prolonged freezing rain events: (**a**) the Scandinavian pattern, (**b**) the East European pattern, (**c**) lee cyclones corresponding to the Scandinavian pattern and (**d**) lee cyclones corresponding to East European pattern. Mean sea level pressure (black contours) and 2 m temperature (shaded according to the scale).

Table 1. Classification of the moderate (2 consecutive days) and long-duration (3 consecutive days) freezing rain events in Bucharest based on the origin of the high-pressure system. The event analyzed in this article is indicated in red.

No.	Interval	No. of Days with Freezing Rain	Classification of High-Pressure System
1	30 Dec–1 Jan 1986	3	Scandinavian origin
2	13–15 Jan 1987	3	Scandinavian origin
3	1–3 Dec 1988	3	Scandinavian origin
4	24–26 Jan 2019	3	Scandinavian origin
5	6–7 Jan 1985	2	Scandinavian origin
6	9–10 Feb 1985	2	East-European origin
7	20–21 Nov 1985	2	East-European origin
8	7–8 Feb 1987	2	East-European origin
9	21–22 Nov 1993	2	East-European origin
10	27–28 Dec 1995	2	Scandinavian origin
11	4–5 Feb 1996	2	Scandinavian origin
12	2–3 Jan 1997	2	Scandinavian origin
13	11–12 Jan 1997	2	Scandinavian origin
14	4–5 Dec 1998	2	East-European origin
15	26–27 Dec 1999	2	Lothar windstorm periphery
16	6—7 Dec 2002	2	East-European origin
17	1—2 Dec 2010	2	Scandinavian origin
18	4—5 Feb 2012	2	Scandinavian origin
19	29–30 Nov 2014	2	East-European origin
20	7—8 Feb 2017	2	Scandinavian origin

5. Conclusions

An exceptional freezing rain event, that occurred between 24–26 January 2019 in Bucharest, Romania was analysed. Based on weather observations from Bucharest, a climatology of freezing rain between 1980–2018 was constructed. From a total of 59 periods with consecutive days with freezing rain, 27% of all periods were moderate-duration events (2 consecutive days) and only 3% were long-duration events (3 consecutive days).

The meteorological analysis of the moderate and long-duration events revealed a particular sub-synoptic pattern that involves a high-pressure system North of Romania and a low-pressure system South of Romania. While the low-pressure system has always a Mediterranean origin, the high-pressure system can have multiple origins: from Scandinavian Peninsula (favouring the transport of arctic air masses) and from Eastern Europe (favouring the transport of polar air masses). The highest incidence of moderate and long-duration events involve a Scandinavian origin anticyclone. A particular freezing rain event was identified during 26–27 December 1999, involving a cyclonic pattern of North-Atlantic origin. This case requires further analyses as it is related with the *Lothar* windstorm.

Due to the distribution of high and low-pressure values induced by the local topography, the Carpathians lee cyclogenesis is triggered. Consequently, surface cold air mass uplifted the pre-existent warm air to the mid-level. Generally, the warm core is visible within the middle layers for between 12 and 24 h. If this process coincides with the advection of warm and humid air from the Mediterranean Sea, the precipitation episode is enhanced and its duration increases. The analysed case from 24–26 January 2019 was the most prolonged event registered so far, with 33 hourly reports of freezing rain in Bucharest. The high-pressure system involved had Scandinavian origin.

Because the database of freezing rain events has been built only for Bucharest city and surroundings, further in-depth analysis for a wide area will make the subject of a subsequent paper, focus on understanding the mesoscale processes and the role of Carpathian Mountains.

Author Contributions: Conceptualization, S.A., D.E. and B.A.; methodology, S.A., M.B. and D.E.; resources, L.M. and J.V.; software, D.E. and M.B.; formal analysis, S.A. and B.A.; investigation, S.A., D.E. and B.A.; writing—original draft preparation, S.A. and C.A.M.; writing-all authors; writing—review and editing, D.E., L.M. and J.V.; visualization, D.E., S.A. and M.B.; supervision, S.A. and D.E.

Funding: This research was funded by the Ministry of Research and Innovation through Program I–Development of the national research development system, Subprogram 1.2–Institutional Performance–Projects of Excellence Financing in RDI, Contract No.19PFE/17.10.2018, by the Romanian National Core Program Contract No.18N/2019; and by the European Regional Development Fund through the Competitiveness Operational Programme 2014–2020, POC-A.1-A.1.1.1-F-2015, project Research Centre for environment and Earth Observation CEO-Terra, SMIS code 108109, contract No. 152/2016.

Acknowledgments: The authors acknowledge the Copernicus–European Union's Earth Observation Programme Service for providing ERA5 data, Copernicus Climate Change Service Climate Data Store, the University of Wyoming for maintaining the sounding database and the Romanian Emergency Service for providing the statics of emergency calls related to the freezing rain event analyzed in the article. The photo showed in Figure 2a was provided by Simona Andrei and the photos in Figure 2b,c were provided by Luminiţa Mărmureanu.

References

1. American Meteorological Society. Freezing Rain. *Gloss. Meteorol.* **2019**. Available online: http://glossary. ametsoc.org/wiki/Freezing_rain (accessed on 1 November 2019).
2. Carriére, J.M.; Lainard, C.; Bot, C.L.; Robart, F. A climatological study of surface freezing precipitation in Europe. *Meteorol. Appl.* **2000**, *7*, 229–238. [CrossRef]
3. Kämäräinen, M.; Hyvärinen, O.; Jylhä, K.; Vajda, A.; Neiglick, S.; Nuottokari, J.; Gregow, H. A method to estimate freezing rain climatology from ERA-Interim reanalysis over Europe. *Nat. Hazards Earth Syst. Sci.* **2017**, *17*, 243–259. [CrossRef]
4. Martner, B.E.; Snider, J.B.; Zamora, R.J.; Byrd, G.P.; Niziol, T.A.; Joe, P.I. A remote-sensing view of a freezing-rain storm. *Mon. Wea. Rev.* **1993**, *121*, 2562–2577. [CrossRef]

5. Norrman, J.; Eriksson, M.; Lindqvist, S. Relationships between road slipperiness, traffic accident risk and winter road maintenance activity. *Clim. Res.* **2000**, *15*, 185–193. [CrossRef]

6. Rauber, R.; Olthoff, L.; Ramamurthy, M.; Miller, D.; Kunkel, K. A Synoptic Weather Pattern and Sounding-Based Climatology of Freezing Precipitation in the United States East of the Rocky Mountains. *J. Appl. Meteor.* **2001**, *40*, 1724–1747. [CrossRef]

7. Juga, I.; Hippi, M.; Moisseev, D.; Saltikoff, E. Analysis of weather factors responsible for the traffic 'Black Day' in Helsinki, Finland, on 17 March 2005. *Meteorol. Appl.* **2012**, *19*, 1–9. [CrossRef]

8. Degelia, S.; Christian, J.; Basara, J.; Mitchell, T.; Gardner, D.; Jackson, S.; Raglanda, J.; Mahana, H. An overview of ice storms and their impact in the UnitedStates. *Int. J. Climatol.* **2015**, *36*, 2811–2822. [CrossRef]

9. Brooks, C. The nature of sleet and how it is formed. *Mon. Wea. Rev.* **1920**, *48*, 69–73. [CrossRef]

10. Penn, S. The prediction of snow vs. rain. Forecasting Guide No. 2. In *U.S. Weather Bureau*; NOAA Central Library: Silver Spring, MD, USA 1957; p. 29.

11. Mahaffy, F. The Ice Storm of 25–26 February 1961 at Montreal. *Weatherwise* **1961**, *14*, 241–244. [CrossRef]

12. Stewart, R.; King, P. Freezing Precipitation in Winter Storms. *Mon. Weather Rev.* **1987**, *115*, 1270–1279. [CrossRef]

13. Pratert, E.; Borho, A. Doppler Radar Wind and Reflectivity Signatures with Overrunning and Freezing-Rain Episodes: Preliminary Results. *J. Appl. Meteorol.* **1992**, *31*, 1350–1360. [CrossRef]

14. Rauber, R.M.; Ramamurthy, M.K.; Tokay, A. Synoptic and mesoscale structure of a severe freezing rain event: The St. Valentine's Day ice storm. *Weather Forecast.* **1994**, *9*, 183–208. [CrossRef]

15. Thoreson, A.D. Doppler radar case study of a freezing rain event. In Proceedings of the 27th Conference on Radar Meteorology, Vail, CO, USA, 9–13 October 1995, pp. 412–413.

16. Zerr, R. Freezing Rain: An Observational and Theoretical Study. *J. Appl. Meteorol.* **1997**, *36*, 1647–1661. [CrossRef]

17. Cortinas, J., Jr. A climatology of freezing rain over the Great Lakes region of North America. *Mon. Weather Rev.* **2000**, *128*, 3574–3588. [CrossRef]

18. Kwon, S.; Byun, H.R.; Park, C.; Kwon, H.N. On the Freezing Precipitation in Korea and the Basic Schemes for Its Potential Prediction. *Asia–Pac. J. Atmos. Sci.* **2016**, *52*, 35–50. [CrossRef]

19. Bocchieri, J. The Objective Use of Upper Air Soundings to Specify Precipitation Type. *Mon. Weather Rev.* **1980**, *108*, 596–603. [CrossRef]

20. Huffman, G.; Norman, G., Jr. The Supercooled Warm Rain Process and the Specification of Freezing Precipitation. *Mon. Weather Rev.* **1988**, *116*, 2172–2182. [CrossRef]

21. Rauber, R.; Olthoff, L.S.; Ramamurthy, M.K.; Kunkel, K.E. The relative importance of warm rain and melting processes in freezing precipitation events. *J. Appl. Meteor.* **2000**, *39*, 1185–1195. [CrossRef]

22. McCray, C.; Atallah, E.; Gyakum, J. Long-Duration Freezing Rain Events over North America: Regional Climatology and Thermodynamic Evolution. *Weather Forecast.* **2019**, *34*, 665–681. [CrossRef]

23. Ressler, G.; Milrad, S.; Atallah, E.; Gyakum, J. Synoptic-Scale Analysis of Freezing Rain Events in Montreal, Quebec, Canada. *Weather Forecast.* **2012**, *27*, 362–378. [CrossRef]

24. Park, C.; Byun, H.R. Three Cases with the Multiple Occurrences of Freezing Rain in One Day in Korea (12 January 2006; 11 January 2008; and 22 February 2009). *Atmosphere* **2015**, *25*, 31–49. [CrossRef]

25. Bresson, E.; Laprise, R.; Paquin, D.; Thériault, J.; de Elía, R. Evaluating the ability of CRCM5 to simulate mixed precipitation. *Atmos. Ocean* **2017**, *52*, 79–93. [CrossRef]

26. Razy, A.; Milrad, S.M.; Atallah, E.H.; Gyakum, J.R. Synoptic-scale environments conducive to orographic impacts on cold-season surface wind regimes at Montreal, Quebec. *J. Appl. Meteorol. Climatol.* **2017**, *51*, 598–616. [CrossRef]

27. Bernstein, B. Regional and local influences on freezing drizzle, freezing rain, and ice pellets. *Weather Forecast.* **2000**, *15*, 485–508. [CrossRef]

28. Robbins, C.C.; Cortinas, J.V., Jr. Local and Synoptic Environments Associated with Freezing Rain in the Contiguous United States. *Weather Forecast.* **2002**, *17*, 47–65. [CrossRef]

29. Markosek, J. Severe freezing rain in Slovenia. *Eur. Forecast. Newsl.* **2015**, *20*, 38–42. [CrossRef]

30. Vajda, A. Impacts of severe winter weather events on critical infrastructure. In Proceedings of the RAIN WP2 Workshop on Past Severe Weather Hazards, Berlin, Germany, 27 February 2015. Available online: http://rain-project.eu/wp-content/uploads/2015/03/Vajda_RAIN_Berlin_workshop_FMI_AVajda.pdf (accessed on 1 November 2019).

31. Manea, A.; Raliţă, I.; Dumitrescu, A.; Sommerfeld, A.; Wichura, B. Analysis of spatial and temporal distribution of freezing rain events in Romania and Germany. In Proceedings of the 13th International Workshop on Atmospheric Icing of Structures (IWAIS 2009), Andermatt, Switzerland, 8–11 September 2009.

32. Eurostat. Population on 1 January by Age Groups and Sex-Cities and Greater Cities. 2019. Available online: http://appsso.eurostat.ec.europa.eu/nui/show.do?dataset=urb_cpop1&lang=en (accessed on 11 February 2019).

33. Hersbach, H.; Bell, B.; Berrisford, P.; Horányi, A.; Sabater, J.; Nicolas, J.; Radu, R.; Schepers, D.; Simmons, A.; Soci, C.; et al. Global reanalysis: Goodbye ERA-Interim, hello ERA5. *ECMWF Newsl.* **2019**, *159*, 17–24. [CrossRef]

34. Peters, G.; Fischer, B.; Andersson, T. Rain observations with a vertically looking Micro Rain Radar (MRR). *Boreal Environ. Res.* **2002**, *7*, 353–362.

35. Kneifel, S.; Maahn, M.; Peters, G.; Simmer, C. Observation of snowfall with a low-power FM-CW K-band radar (Micro Rain Radar). *Meteorol. Atmos. Phys.* **2011**, *113*, 75–87. [CrossRef]

36. Cha, J.; Chang, K.; Yum, S.; Choi, Y. Comparison of the bright band characteristics measured by Micro Rain Radar (MRR) at a mountain and a coastal site in South Korea. *Adv. Atmos. Sci.* **2009**, *26*, 211–221. [CrossRef]

37. Weather Graphics. Global Surface Archives [World Meteorological Organization FM-12 SYNOP and FM-15 METAR Coded Data File. 2019. Available online: http://www.weathergraphics.com/gsa (accessed on 1 November 2019).

38. Cortinas, J., Jr.; Bernstein, B.; Robbins, C.; Strapp, J. An Analysis of Freezing Rain, Freezing Drizzle, and Ice Pellets across the United States and Canada: 1976–90. *Weather Forecast.* **2004**, *19*, 377–390. [CrossRef]

39. Reeves, H. The uncertainty of precipitation-type observations and its effect on the validation of forecast precipitation type. *Weather Forecast.* **2016**, *31*, 1961–1971. [CrossRef]

40. Micu, D.; Dumitrescu, A.; Cheval, S.; Bîrsan, M. *Climate of the Romanian Carpathians*; Springer International Publishing: Cham Heidelberg, Germany; New York, NY, USA; Dordrecht, The Netherlands; London, UK, 2015; p. 213. [CrossRef]

41. Bordei-Ion, E. *Rolul Lantului Alpino-Carpatic in Evolutia Ciclonilor Mediteraneeni (The Role of the Alpine-Carpathic Chain Related to the Evolution of the Mediterranean Cyclones)*, 2nd ed.; Printech: Bucureşti, Romania, 2009; p. 138.

42. Reeves, H.; Laaksonen, A.; Alper, M.E. Increasing large scale windstorm damage in Western, Central and Northern European forests, 1951–2010. *Sci. Rep.* **2017**, *7*, 46397. [CrossRef]

Article

Estimation of CO_2 Emissions from Wildfires Using OCO-2 Data

Meng Guo [1], Jing Li [2],*, Lixiang Wen [1] and Shubo Huang [1]

[1]   Key Laboratory of Geographical Processes and Ecological Security in Changbai Mountains, Ministry of Education, School of Geographical Sciences, Northeast Normal University, Changchun 130024, China; guom521@nenu.edu.cn (M.G.); wenlx679@nenu.edu.cn (L.W.); huangsb966@nenu.edu.cn (S.H.)

[2]   Northeast Institute of Geography and Agricultural Ecology, Chinese Academy of Science, Changchun 130102, China

*   Correspondence: lijingsara@iga.ac.cn

Received: 15 August 2019; Accepted: 24 September 2019; Published: 25 September 2019

Abstract: The biomass burning model (BBM) has been the most widely used method for estimation of trace gas emissions. Due to the difficulty and variability in obtaining various necessary parameters of BBM, a new method is needed to quickly and accurately calculate the trace gas emissions from wildfires. Here, we used satellite data from the Orbiting Carbon Observatory-2 (OCO-2) to calculate CO_2 emissions from wildfires (the OCO-2 model). Four active wildfires in Siberia were selected in which OCO-2 points intersecting with smoke plumes identified by Aqua MODIS (MODerate-resolution Imaging Spectroradiometer) images. MODIS band 8, band 21 and MISR (Multi-angle Imaging SpectroRadiometer) data were used to identify the smoke plume area, burned area and smoke plume height, respectively. By contrast with BBM, which calculates CO_2 emissions based on the bottom–top mode, the OCO-2 model estimates CO_2 emissions based on the top–bottom mode. We used a linear regression model to compute CO_2 concentration (XCO_2) for each smoke plume pixel and then calculated CO_2 emissions for each wildfire point. The CO_2 mass of each smoke plume pixel was added to obtain the CO_2 emissions from wildfires. After verifying our results with the BBM, we found that the biases were between 25.76% and 157.11% for the four active fires. The OCO-2 model displays the advantages of remote-sensing technology and is a useful tool for fire-emission monitoring, although we note some of its disadvantages. This study proposed a new perspective to estimate CO_2 emissions from wildfire and effectively expands the applied range of OCO-2 satellite data.

Keywords: wildfire; CO_2 emission; OCO-2; MISR; MINX

1. Introduction

Land carbon storage is strongly influenced by ecosystem disturbances, which influence species composition and structure and cause a net carbon stock reduction [1]. Wildfires are one of the most significant disturbances to land ecosystems on regional and global scales and generate large amounts of critical greenhouse gases (GHG), especially CO_2. Wildfires play a key role in global biogeochemical cycles, atmospheric composition, and land ecosystem attributes, all of which influence climate systems [1,2]. Biomass burning events in forests have released approximately 2–3 Pg C yr^{-1}, mostly in the form of CO_2 [2–5]. With increasing temperatures due to global warming and the accompanying drying trends, wildfires have increased in boreal forests over the last few decades [6].

Boreal forests are the most important global terrestrial carbon stocks, and CO_2 released into the atmosphere by wildfires may convert forests from carbon sinks into net sources, which in turn contributes to global warming and affects the carbon cycle [7]. Wildfires also impact the climate through changes in the surface albedo [8]. Siberia's boreal forests cover more than 60% of the total global area [7]. The ability to accurately estimate CO_2 emissions from boreal forest fires is necessary to understand the role of Siberia in the global carbon cycle.

Seiler and Crutzen [9] were the first to propose a biomass burning model (BBM) to calculate CO_2 released from biomass burning. Following that, many researchers have estimated CO_2 emissions from wildfires in different regions using the BBM. Miranda, et al. [10] forecasted CO_2 emissions from forest fires in Portugal between 1988 and 2010 and argued that CO_2 emissions increased from 28 Tg in 1988 to 44 Tg in 2010. Wiedinmyer and Neff [11] calculated CO_2 emissions from fires in the US and found ~80 Tg of CO_2 yr^{-1} were emitted in Alaska from 2002 to 2006. With the help of remote-sensing imagery, burned area data were retrieved and combined with results from previous research on burning efficiency and emission factors, showing that annual forest fire CO_2 emissions from 23 European countries were 8.4–20.4 Tg [12]. Werf, et al. [13] estimated the fire emissions from 1997 to 2009 with 0.5-degree spatial resolution and monthly temporal resolution using a revised biogeochemical model and found that the average global fire carbon emission was 2.0 Pg C y^{-1} and the contribution from forest fires was 15%. Rosa, et al. [14] studied changes in burned areas and found that CO_2 emission from biomass burning in Portugal was 159–5655 Gg from 1990 to 2008. Shi and Yamaguchi [15] developed a new high-resolution biomass burning emission inventory during 2001–2010 in Southeast Asia based on satellite products and combustion factors and found an average of 817.81 Tg of CO_2 emissions every year. Zhou, et al. [16] calculated the biomass burning emissions in mainland China in 2012 using satellite data and source-specific emission factors and found that 675.30 Tg of CO_2 was emitted.

As we know, many parameters are needed in BBM, such as above-ground fuel load, combustion efficiency, and emission factors. However, all of these parameters are not easy to obtain, need field surveys and laboratory tests, and require a large amount of cost for labor and resources. With the development of remote sensing, a new method is needed to improve the efficiency of wildfire emissions.

In recent years, the relationships between of fire radiative power (FRP) with biomass consumed and emission factors have been successfully used to estimate trace gases released into the atmosphere [17–22]. Konovalov, et al. [21] estimated CO_2 emissions from wildfires in Siberia in 2012 using the direct relationship between biomass burning rate (BBR) and FRP and obtained the result of 262–477 Tg C, but the accuracy requires further analysis. Today, many biomass burning emission inventories exist, such as the Global Fire Emissions Database (GFED) [23], Fire INventory from National Center for Atmospheric Research (NCAR) (FINN) [24], and Global Fire Assimilation System (GFAS) [25]. All these data sets have made great contributions to wildfire emission research, but all these products are based on the BBM or FRP. Further, these products constitute global emissions at coarse spatial resolution on monthly or daily temporal scales. No product is available that is focused on biomass burning emissions of active fires or in times of fire occurrence.

Due to the ability of remote-sensing technology to obtain data and repeat sampling in large forested regions, active burning can be identified in inaccessible locations or where using aircraft observation is too costly [26]. It could even be predicted that in the near future, greenhouse gas-monitoring satellites such as GOSAT (Greenhouse Gases Observing Satellite), OCO-2 (Orbiting Carbon Observatory-2), and TanSat (CarbonSat) will play a key role in tracking biomass burning emissions, which will help us better understand the contributions of biomass burning to regional climate change and atmospheric composition [7,27,28]. By using OCO-2 data, Heymann, et al. [27] estimated that CO_2 emissions from Indonesian wildfires in 2015 were 30% lower than provided by the emission databases of GFAS version 1.2 and GFED version 4s. In their paper, they used OCO-2 XCO_2 (column-averaged dry air mole fraction of CO_2) as the main data source and derived the difference between the fire-affected XCO_2 values and background XCO_2 values. Proposing a new method and a new perspective to estimate CO_2 emissions from wildfires. Patra, et al. [28] also used OCO-2 and atmospheric chemistry-transport model to estimate the C release during the 2014~2016 El Niño and found that 2.4 ± 0.2 Pg C was released into the atmosphere.

In our previous work, Guo, et al. [7], we demonstrated for the first time the ability of GOSAT data to estimate CO_2 emissions from forest fires. Given the limitations of GOSAT data in estimating

CO_2 emissions such as sparse observed points of TANSO-FTS (Thermal And Near-infrared Sensor for carbon Observation–Fourier Transform Spectrometer) and a narrow swath of CAI (Cloud and Aerosol Imager) observations mentioned in Guo, et al. [7], this paper attempts to use another GHG satellite, OCO-2, combined with MODIS (MODerate-resolution Imaging Spectroradiometer) and MISR (Multi-angle Imaging SpectroRadiometer) data to estimate CO_2 emissions from wildfires.

2. Materials and Methods

2.1. Orbiting Carbon Observatory-2 (OCO-2) Data

OCO-2, launched on 2 July 2014, is a National Aeronautics and Space Administration (NASA) mission to study CO_2 in the atmosphere to better understand the carbon cycle and to track and quantify CO_2 sources and sinks on a global scale [29] as well as their impact on climate. OCO-2 is a polar sun-synchronous orbit satellite with a return cycle of 16 days and crosses the equator at 13:35 LT (local time). OCO-2 estimates the column-averaged dry air mole fraction of CO_2 (denoted as XCO_2, in ppm) from Earth's surface to the top of the atmosphere. When radiances pass through the atmosphere, CO_2 and O_2 will absorb radiance at specific wavelengths. XCO_2 is calculated by using the ratio of column CO_2 densities and O_2 molecules along the optical path between the Sun, ground surface, and OCO-2 sensor and is then multiplied with the column-averaged O_2 concentration [30–32]. OCO-2 carries a three-band imaging instrument that measures the O_2 A-band at 0.76 μm and two CO_2 bands at 1.61 and 2.06 μm to measure the weak and strong CO_2 bands, respectively [33–36].

OCO-2 is an excellent remote-sensing satellite for the study of atmospheric CO_2 which collects global atmospheric CO_2 with high precision and resolution [27,37,38]. The OCO-2 instrument covers a 1.29×2.25 km footprint at nadir and is capable of acquiring eight cross-track footprints, creating a swath width of 10.3 kilometers, to obtain XCO_2 values. Near-simultaneous measurements of XCO_2 using ground-based the Fourier Transform Spectrometers in Total Carbon Column Observing Network (TCCON) were used to validate the accuracy of XCO_2 products from OCO-2 and found that after bias correction, the XCO_2 extracted from OCO-2 agreed well on a global scale with the TCCON for nadir, glint, and target observations, with median differences less than 0.5 ppm and root-mean-square differences obviously below 1.5 ppm [27,30]. Many researchers found that net variable errors of OCO-2 XCO_2 were ~0.5 to 2.0 ppm over land [29,31].

As we know, OCO-2 only collects data in clear skies because the presence of aerosol, such as clouds, may increase the uncertainty of the XCO_2 value [39]. However, in this study, the bias was ignored as we selected the OCO-2 points which intersected light smoke plumes. We assumed that the bias compared with the clear sky condition was the result of forest fire CO_2 release [7,40,41].

OCO-2 L2 Lite FP V8r data with NetCDF format (downloaded from the websites https://mirador. gsfc.nasa.gov/) which contain daily data of retrieved values for the state vector and geolocation information were used in this study.

2.2. Smoke Plumes and Burned Area Monitor Using Moderate-Resolution Imaging Spectroradiometer (MODIS) Data

MODIS is a key instrument aboard the Terra (across the equator in the morning) and Aqua (across the equator in the afternoon) satellites. Terra MODIS and Aqua MODIS sample the entirety of the Earth's surface with a 2330-km swath, providing near-global daily coverage. MODIS acquires data in 36 spectral bands from 0.4 to 14.4 μm at three spatial resolutions: 250 m (band 1 and 2), 500 m (band 3–7), and 1000-m (band 8–36) covering visible, near-infrared, shortwave infrared (IR), and thermal-IR regions in the electro-magnetic spectrum [42]. Bands 1 to 7 are specifically designed for remote sensing of land [43].

Comparing MODIS false-color images with each single band, we found that MODIS band 8 (4.05–4.2 μm) could identify smoke plumes clearly and has higher correlation with XCO_2 values in the intersected points. In this study, we used MODIS band 8 to identify smoke plumes.

The launch of MODIS effectively improved the capabilities in fire detection compared with other sensors given active fire detection bands that allow for the retrieval of sub-pixel temperatures at 3.9 and 11 μm which have been successfully used for detecting fire locations [42,44,45]. Both actively burning fires and post-fire burned areas can be identified using MODIS data. Middle-infrared-to-thermal wavelengths (3.6–12 μm) were mainly used to detect actively burning fires based on the combustion temperature. The post-fire burned area was mainly extracted visually to a near-infrared band (0.4–1.3 μm) by way of the reduction of surface reflectance when vegetation is removed [46]. MODIS band 21 (3.929–3.989 μm) has been widely used in forest fire and volcano research [47,48], and in this study, we used it to monitor burned areas. Active fires and the recently burned scars all have higher temperatures which can be captured by MODIS band 21.

The orbit of Aqua MODIS is sun-synchronous, near-polar, and crosses the equator at 1:30 p.m. MODIS data and products have been successfully used in many fields, including land cover changes, environmental monitoring, global warming, landscape ecology, etc. In this paper, Aqua MODIS L1b data were used to map the burned scars and to detect smoke plumes.

2.3. Multi-Angle Imaging Spectroradiometer (MISR) Data and MISR Interactive Explorer (MINX) Software

MISR is part of NASA's first Earth Observing System spacecraft, the Terra spacecraft, which was launched on 18 December 1999. Unlike other satellite instruments, MISR records images of the Earth at 9 different angles in each of 4 color bands simultaneously. MISR's 36 simultaneous spectral-angular images have been widely used in aerosol optical depth, cloud, and smoke plumes study [4,49–51].

The MISR INteractive eXplorer (MINX) software is used to retrieve and analyze the heights and winds locally of smoke, dust, and clouds at higher spatial resolution and with greater precision [50,52]. MINX is specially designed to display images from MISR's nine cameras and to determine the height and speed of motion of aerosol plumes and clouds on those images. This software has been the best-established tool in smoke plume height estimation and using this tool, thousands of plume heights have been derived in MISR Plume Height Project (https://misr.jpl.nasa.gov/getData/accessData/ MisrMinxPlumes2/). MINX can use MODIS thermal anomalies to locate active fires, and then computes the smoke plume heights from MISR stereo imagery [51,52]. It is a robust tool in analyzing the physical properties of smoke plumes and plume dynamics. In this study, MINX V4.0 was used to identify smoke plumes and derive the height [52–54].

2.4. Wildfire Cases Selected in Boreal Forest

We targeted wildfires covered by Aqua MODIS images and OCO-2 on the same day where OCO-2 points must intersect wildfire smoke plumes. The MODIS products of MCD64A1 and Landsat images were used to obtain wildfire information. Then, we downloaded the Aqua MODIS L1B images covering the wildfires and identified smoke plumes visually. Based on the time span of wildfires and spatial range of smoke plumes, OCO-2 L2 data covering smoke plumes were downloaded to look for the intersect points. We selected one wildfire that met our standard in Siberia (on 20 April 2015), the cross time was 20 April 2015 05:13:08 and 20 April 2015 05:15:00 LT for OCO-2 and Aqua MODIS, respectively (Figure 1).

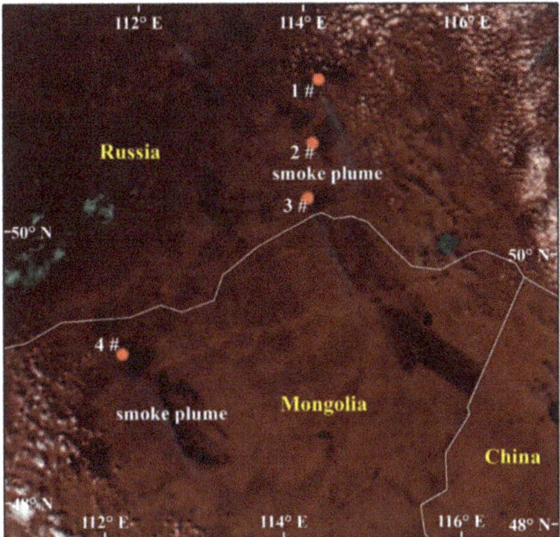

Figure 1. Moderate-resolution Imaging Spectroradiometer (MODIS) false-color image (red-green-blue (RGB) = 741) of study region. The white lines indicate national boundaries and the red point is fire point. Smoke plumes can be identified from the image. We found 3 active fires in Russia and 1 in Mongolia, but here, the study region is labeled "Siberia".

2.5. CO_2 Emission Method (OCO-2 Model)

In this study, we are going to explore a new method to quantify the CO_2 released from wildfires that is based on the transitional equation of mass calculation as shown below:

$$E_{CO2} = \rho \times V \tag{1}$$

E_{CO2} indicates CO_2 emissions from wildfire, ρ is the density of CO_2, and V is the volume of smoke plumes calculated using S (area) and multiplied by H (height). S is the area of smoke plumes calculated from MODIS images and H is the height of smoke plumes, which will be derived using MINX. The equation can thus be expressed as follows:

$$E_{CO2} = \rho \times S \times H \tag{2}$$

As shown before, the unit of OCO-2 XCO_2 data is parts per million (ppm). To calculate the quantity of CO_2 released from wildfires, it was necessary to convert these numbers to density values using

$$\rho = M/22.4 \times ppm \times (273/T) \times (Ba/1013.25) \tag{3}$$

where ρ (mg/m^3) is the density of CO_2 in the atmosphere, M is the molecular weight of CO_2, 22.4 indicates the molecular mass of atmosphere, *ppm* is the value of XCO_2, T is the temperature (K), and Ba is the atmospheric pressure (hPa) [7]. XCO_2 values are determined by dry air, but M and 22.4 values are within standard atmospheric pressure. In this research, we ignore the uncertainty of water vapor concentration in the atmosphere and the smoke was uniformly distributed from the ground surface to the top of smoke plumes. We also assume that all smoke that released from forest fires was captured by MODIS without being dispersed into the atmosphere. To calculate the CO_2 emissions from wildfires, we must know $\triangle XCO_2$, which is the difference between XCO_2 in smoke plume regions and the background value of XCO_2. Base XCO_2 is represented as the mean value of OCO-2 points surrounding the smoke plumes (~10 km around) and the value is 399.786 ppm.

The temperature and atmospheric pressure are referenced from the L137 model-level definitions (https://www.ecmwf.int/en/forecasts/documentation-and-support/137-model-levels).

2.6. Biomass Burning Model (BBM)

BBM [9,41] has been widely used to evaluate trace gas emission from wildfires and for studying the release of trace gases from biomass burning [55–57]. It is calculated based on the conservation of mass:

$$Es = S \times B \times \beta \times EF \tag{4}$$

where Es (Mg) is CO_2 emissions from wildfires, S is the burned area (ha) as indicated from satellite images, B is the biomass fuel load (Mg ha^{-1}), β is the combustion factor (%), and EF is the emission factor of CO_2 (kg kg^{-1}).

Many researchers used different approaches to obtain the BBM parameter for different ecological systems [13,56,58]. Among them, Akagi, et al. [59] summarized much of the literature data for biomass consumption for different vegetation and fire types, and in this paper we used the parameter of Akagi, et al. [59] to calculate CO_2 emissions which were used to verify the accuracy of OCO-2 model.

3. Results

3.1. Smoke Plume Area Detection and Height Derivation

To calculate CO_2 emissions from wildfires, we must know smoke plume characteristics such as height, distribution, and the value of XCO_2. First, smoke plumes from wildfire must be detected accurately. Remote-sensing data can accomplish this work quite well and MODIS data were suitable for this study. Using the threshold of MODIS band 8, we detected the smoke plumes of the 4 fire cases (Figure 2) with area of 2.48×10^3 km^2, 2.41×10^3 km^2, 2.50×10^3 km^2 and 6.48×10^3 km^2, respectively.

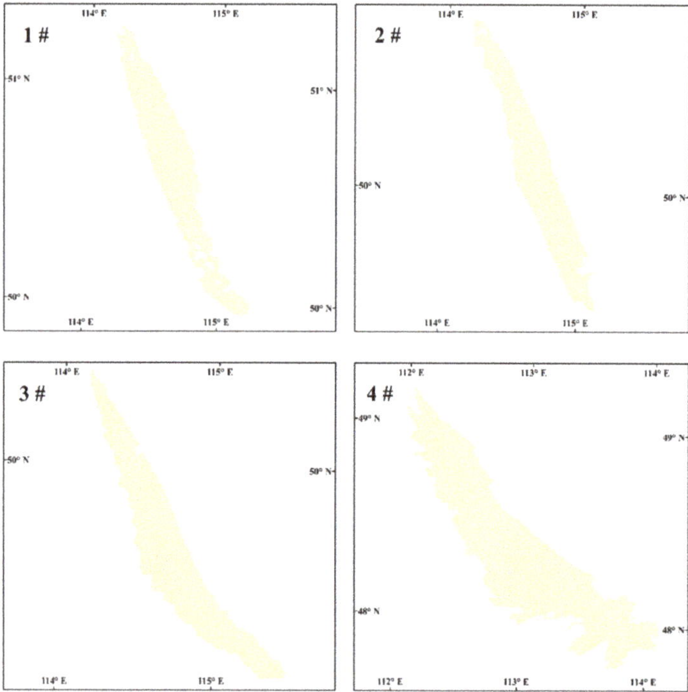

Figure 2. Smoke plumes detection using MODIS band 8.

As mentioned above, MINX has the potential to extract smoke characteristics and here we combine MINX and MODIS data to identify smoke plumes. By employing MINX v4.0, we derived the height of each wildfire smoke plume (Figure 3). We calculated the average height of each smoke plume pixel and obtained the height value as 2.29 km, 2.38 km, 1.99 km, and 2.61 km for the 4 smoke plumes, respectively.

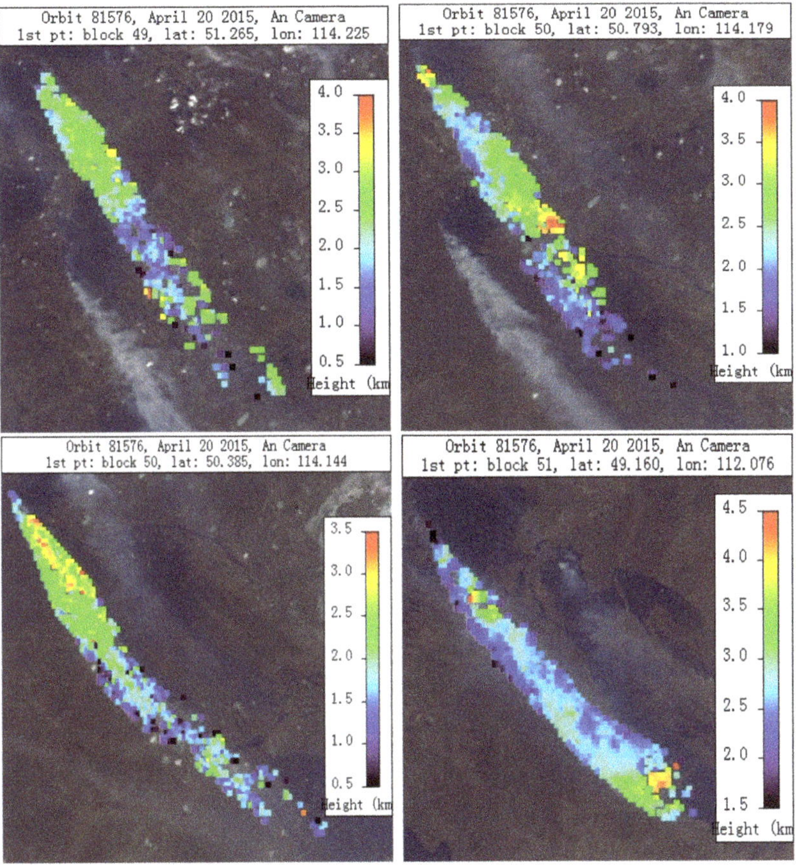

Figure 3. Height of smoke plume sketches derived from Multi-Angle Imaging Spectroradiometer (MISR) data by employing MISR Interactive Explorer (MINX) v4.0 software.

From MINX results we can also obtain the smoke plume area of 1.46×10^3 km^2, 1.31×10^3 km^2, 1.36×10^3 km^2, and 1.04×10^3 km^2, respectively which are rather lower than that from MODIS band 8. MINX was robust in deriving aerosol height but could only recognize thick smoke which could be easily identified from true color images. MODIS band 8 has the potential to derive thin smokes using the reflectance threshold.

3.2. Burned Area Monitor

Burned area of forest fires could be accurately and rapidly monitored using remote-sensing images based on sharp differences in the spectral characteristics of the surface inside and outside of the burned scars. Using MODIS band 21 (MYD021KM.A2015110.0515.006.2015110190452) fire points were observed and visual interpretation was used to map the burned scars (Figure 4). Here, we assume that the smoke plumes were released from the mapped burned scars. Given that the spatial resolution of MODIS is 1000 m, the burned scars must contain many mixed pixels. This means that the burned areas used in this study are worth discussion.

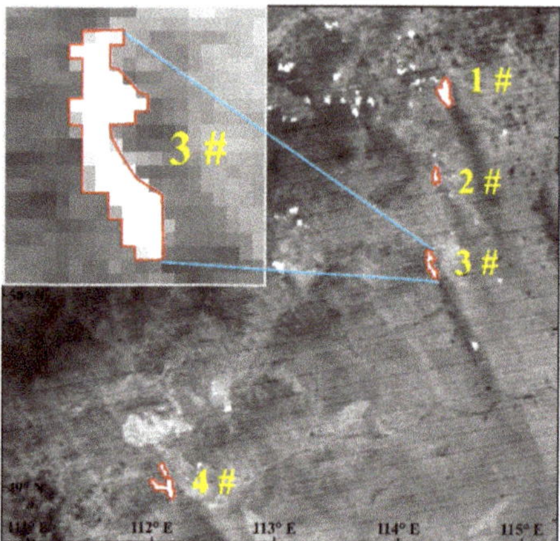

Figure 4. Burned area from MODIS band 21, and the upper-left corner is the enlarged images of the 3# fire scar.

3.3. Model XCO_2 from Smoke Plumes

The XCO_2 values of smoke plume pixels changed with the smoke concentration. To calculate CO_2 emissions accurately, we first calculated XCO_2 values for each smoke plume pixel and then subtracted clear sky XCO_2 values. In this paper, linear regression analysis was used based on OCO-2 XCO_2 values and MODIS band 8 reflectance in the intersection regions. Scatter plots of XCO_2 against MODIS band 8 reflectance in the study region are shown in Figure 4 and close correlations were found ($R^2 = 0.655$, $p < 0.001$). By using the relationships between MODIS band 8 reflectance and XCO_2, we could calculate XCO_2 of the smoke plume pixels that did not intersect with OCO-2 points (Figure 5). In Figure 6 we see that XCO_2 also has different values because of differences in smoke concentration. Higher values are in the center, and lower values are at the edges of the smoke plumes. XCO_2 values decreased with increasing transport distance.

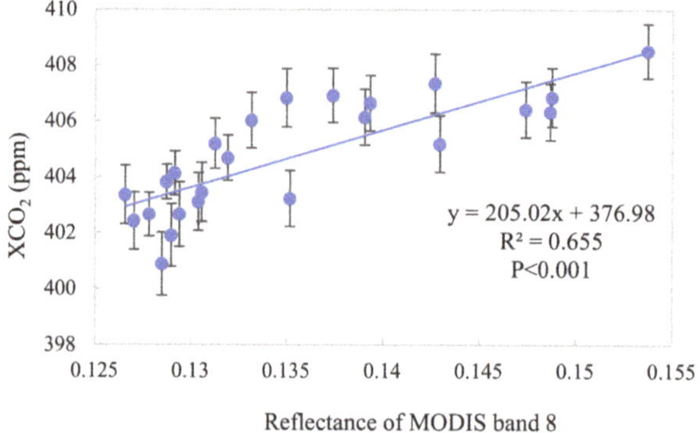

Figure 5. Scatter plots of reflectance of MODIS band 8 vs. Orbiting Carbon Observatory-2 (OCO-2) CO_2 concentration in the study region.

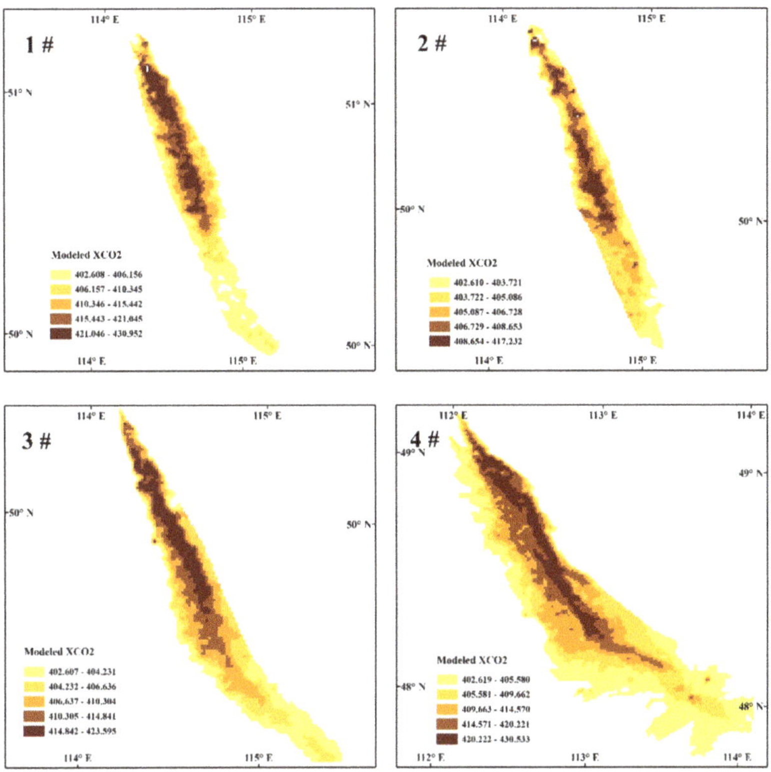

Figure 6. XCO2 values of smoke plumes calculated from MODIS band 8 reflectance.

3.4. CO₂ Emission Calculation from OCO-2 Model

3.4. CO_2 Emission Calculation from OCO-2 Model

Equations 1–3 (named the OCO-2 model) were used to calculate CO_2 emissions from wildfires. In the OCO-2 model, the height of the smoke plumes was derived from MISR data using MINX v4.0 software, the smoke plumes were derived from MODIS band 8. The calculated CO_2 emissions for each fire point are shown in Table 1. We found that CO_2 emissions were from 122.21 (±4.87) × 10^3 Mg for the 2# wildfire to 288.70 (±14.33) × 10^3 Mg for the 4# wildfire.

Table 1. CO_2 emission values and calculated parameters for each fire point by OCO-2.

Fire Point No.	Smoke Plume Area (10^3 km²)	Smoke Plume Height (km)	Base XCO₂ (ppm)	T (K)	Ba (hPa)	CO₂ Emission (10^3 Mg)
1#	2.48	2.29				156.25 (±4.85)
2#	2.41	2.38	399.786	281.59	906.97	122.21 (±4.87)
3#	2.50	1.99				122.84 (±4.21)
4#	6.48	2.61				288.70 (±14.33)

4. Discussion

4.1. Smoke Plume Height Estimation

Energy from wildfires, especially crown fire, can loft smoke plumes above the boundary layer and promote long-distance transport of gases and particulate matter. Not all smoke aerosols rise high enough and form plumes. A great number of smoke aerosols remain in the near-ground layer [4,49].

From OCO-2 model we know that the height of smoke plume is rather important in determining the accuracy of CO_2 emissions. Regarding the height of smoke plumes, researchers acquire different results using different methods for different fires. Vadrevu, et al. [60] used Cloud-Aerosol Lidar and Infrared Pathfinder Satellite Observations (CALIPSO) to monitor the mean smoke height and suggested that the value was from 2.5 to 9.3 km with an average height of 5.35 km at the northern border of India. Using multi-wavelength lidar, MODIS, and CALIOP imageries, Wu, et al. [61] obtained different elevation values of smoke plumes from 1.5 to 6.5 km for different fire case locations. Ross, et al. [41] assumed the plume height was 0.3, 1, 2, and 5 km to assess the ratio of XCO_2 and XCH_4 from GOSAT in the presence of wildfire plumes. Guo, et al. [7] used GOSAT FTS-TIR L2 data to identify the boundary of smoke plumes in a Russian wildfire of 2010 and obtained the height of 1.58 km; however, this result warrants discussion because of the coarse resolution of GOSAT profile data. Kahn, et al. [50] calculated the height of smoke plumes in Australia using the Multi-angle Imaging SpectroRadiometer (MISR) instrument and found wind-corrected values of approximately 3.5 km. Mazzoni, et al. [49] employed MISR and MODIS data collected over North America during the summer of 2004 and found that the plume height was approximately 3 km (see Figure 5 in Mazzoni, et al. [49]).

By employing MINX v4.0, we calculated the height of each wildfire smoke plume as shown in Figure 3. To simplify, we calculated the average height of each smoke plume pixel and obtained the height value as 2.29 km, 2.38 km, 1.99 km, and 2.61 km respectively for the 4 smoke plumes. Compared with former research, we found that our results were slightly lower for the average smoke plume height which is mainly due to lower energy released from small burned areas.

4.2. CO_2 Emissions by BBM

Using OCO-2 model, we have estimated CO_2 releases from 4 fire scars in Siberia. As we have shown, this is the first time that there has been a determination of CO_2 emissions from forest fires using OCO-2 data. We employed BBM, the most commonly used method in estimating trace gases emission from biomass burning [11,13,21], to verify the accuracy of our results.

As mentioned above, it is difficult and complex to obtain these variables accurately because of the high spatial heterogeneity of vegetation, different weather conditions, and different behaviors of each fire [7,14]. It is also quite difficult for us to field measure all these variables in Siberia by ourselves. Luckily, each variable, such as biomass fuel load, combustion factor, and *EF*, have been discussed by many researchers in recent years. Based on the former researchers' results, Werf, et al. [13] argued that the *EF* in an extratropical forest was 1572 g kg^{-1}. Based on localized measurements, Zhou, et al. [16] found that the *EF* was 1543–1630 g kg^{-1} in Chinese forests. The study by Shi and Yamaguchi [15] found an *EF* of 1580 g kg^{-1} in the forests of Southeast Asia. Chang and Song [56] used an *EF* value of 1618 g kg^{-1} for the forest to estimate biomass burning emissions in tropical Asia. Akagi, et al. [59] used emission ratios (the molar ratio between the two emitted compounds) to derive *EF* based on the carbon mass balance method. Laboratory- and ground-based measurements and airborne measurements of *EF* for burning organic soils, peat, and woody/down/dead vegetation by many researchers were considered in order to calculate *EF* by Akagi, et al. [59], who concluded that the *EF* of CO_2 for boreal forests is 1.489 (±0.121) kg kg^{-1}. The study region in this work is located at high latitudes where the most important boreal forests in the world exist which is similar with Akagi et al.'s research. Thus, in this paper, we used the *EF* of 1.489 (±0.121) kg kg^{-1}.

Akagi, et al. [59] summarized much of the literature data for biomass consumption (biomass loading multiply combustion factor) for different vegetation and fire types and we used their result of 38 Mg ha^{-1} here.

CO_2 emissions from each fire scar and all other variables are shown in Table 2. The comparisons with the results of the OCO-2 model are also listed. We found that the results by the BBM are higher than those of the OCO-2 model for all 4 fire scars. The most similar results were found in the second fire scar (25.76%); the largest bias was in the first fire scar (157.11%) followed by the third (121.10%). We found that all of the study cases have an equal magnitude of CO_2 emission results by the BBM and

OCO-2 models. Thus, we can propose that OCO-2 has great potential in estimating CO_2 emissions from wildfires.

Table 2. CO_2 emission values and calculated parameters for each fire scar using the BBM.

Fire Point No.	Burned Area (ha)	B × β (Mg/ha)	EF (kg/kg)	CO_2 Emission (10^3 Mg)	Compared with OCO-2 Model
1#	7100			401.73 (±32.65)	+157.11%
2#	2700			153.69 (±12.41)	+25.76%
3#	4800	38	1.489 (0.121)	271.59 (±22.07)	+121.10%
4#	7500			424.37 (±34.49)	+46.99%

4.3. CO_2 Emission Differences between the Two Models

Table 2 show the different CO_2 releases from boreal forest fires between BBM and OCO-2 models. We found that the BBM results are higher than those of the OCO-2 (from 25.76% to 157.11%) but remain within our expected range.

The differences may be due to the uncertainty variables of BBM. The biomass fuel load and combustion factors are crucial but difficult to estimate accurately. Biomass fuel load estimates are complex because of the high spatial heterogeneity of the different types of vegetation sampled. Biomass fuel load (here, we just consider the aboveground biomass) was calculated at three levels: litter, surface fuels, and tree crowns; all these three levels change with species and vegetation type and are also affected by the age structure of the forest, the years since the last disturbance, and so on [14,41]. Saatchi, et al. [62] found that the error of forest biomass fuel load is about 50%. The boreal forest is made up by many different biological communities with different biomass loads. The burned scars, especially the larger ones, usually contain many biological communities. The study region is located at the edge of Siberia and the vegetation communities may differ with former research. In this paper, we did not consider biological communities and used one value for the study region. Biomass fuel load based on vegetation community information may increase the estimation accuracy.

Another variable that affected the BBM result is combustion factor, which indicates the ratio of the burned biomass during a fire and is also difficult to estimate because of differences in vegetation characteristics (i.e., plant age, growth cycle, and water content), as well as the unique behavior of each fire. Many researchers have argued that the combustion factor is affected by fire type (surface fire, crown fire, or ground fire), fire phase (flaming or smoldering), and many other factors, such as wind speed, month of occurrence (vegetation has different water content in spring and summer), soil moisture, and even different slope aspects; the different regions ranges of the combustion factor are from 34%~69% with an uncertainty of 50% [7,13,56,59,63]. Uncertainties over the emission factor can also not be ignored and researchers obtained different *EF* values for boreal forests [13,59].

Burned area is also an important factor that affects CO_2 emissions. CO_2 emissions from forest fires have a strong correlation with burned area; accurate measurement of burned area has been crucial for CO_2 emission estimates [14]. Post-fire burned area could be fairly accurately mapped by airborne and satellite imageries of boreal forests but in this paper, the burned areas were not the post-fire areas. Due to the higher reflectance of burning biomass the burned area may be enlarged in this study. We looked at actively burning forest and freshly burned areas that could correspond with smoke plumes, meaning we could not use any other higher spatial-resolution images aside from those of Aqua MODIS.

Uncertainties in the OCO-2 model may come from the smoke plume area, the modeled XCO_2 value of each pixel, as well as the bias of OCO-2 retrieval XCO_2 values. In this study, the smoke plumes were identified by the thresholds of MODIS band 8 which were affected by the researcher's knowledge of remote sensing.

O'Dell, et al. [64] argued that the retrieval XCO_2 will lead to a small positive bias of about 0.3 ppm when thin cirrus appears. They also defined the clouds as the aerosol optical depth (AOD) greater than

0.3. In this study we used MODIS product (MYD04, version 061) with a spatial resolution of 3 km to view the AOD value of wildfire smoke plumes that overlaid with the OCO-2 footprint. We found that all the AOD values of smoke plumes used for the XCO_2 model were lower than 0.3 which means that we could ignore this bias. With the increasing AOD of smoke plumes, the modeled XCO_2 bias may increase; here we did not consider this error. Thermal Infrared Red (TIR) satellite soundings have low sensitivity below 2–3 km of altitude, especially in the boundary layer [65], therefore OCO-2 retrieved XCO_2 may be underestimated in the smoke plumes boundary layer. We also did not consider this bias in this paper.

4.4. Advantages and Limitations of the OCO-2 Model

The OCO-2 model calculates CO_2 emissions using the $\triangle XCO_2$ value and smoke plume characteristics that can be obtained using remote-sensing techniques. It has the advantage of remote-sensing technologies repeat data acquisition, such as being more efficient and saving labor costs.

However, there are limitations for the OCO-2 model. Firstly, there may be an error when smoke plumes area is identified. The threshold of the smoke plume area identified was based on artificial visual interpretations, and the accuracy depended on the knowledge and understanding of smoke plume characteristics by researchers. Although MISR data could be used as a reference, some smoke plumes, especially the smoke released from earlier burning, will be dispersed into the atmosphere and could not be captured by the MODIS sensor. To some extent, this could explain why the estimated CO_2 emissions using the OCO-2 model are lower than BBM.

Secondly, the most serious limitation of our approach is the lack of data. Given the 16-day recycle time of OCO-2, it is not easy to intersect with the smoke plumes, meaning that smoke from small or short-lived wildfires could not be captured by the OCO-2 satellite. It is time consuming to look for OCO-2 points that intersect MODIS smoke plumes on the same day. In this study, we just attempt to explore a simple and accurate method for using remote sensing data to estimate CO_2 emissions from wildfires and in the future, automatic methods for identifying smoke plumes and selecting the intersected OCO-2 points must be developed.

5. Conclusions

Every year, thousands of wildfires occur in boreal forests and release a great amount of CO_2 into the atmosphere. Wildfires transform forests from carbon sinks into sources. Accurate estimation of GHG, especially CO_2 from wildfire in boreal forests, is necessary to understand the role of boreal forests in carbon cycle and global warming. This study explored the feasibility of using OCO-2 and MODIS data to estimate CO_2 emissions from wildfires in boreal forests and we were able to draw a few conclusions.

Firstly, CO_2 emissions could be calculated using OCO-2 data and Aqua MODIS images because of the close passing time over the same territories. MODIS band 8 and 21 could be used in smoke plume identification and burned area monitoring. Because of the higher reflectance of burning biomass, the burned area identified in this study may be enlarged and lead to higher CO_2 emissions from BBM.

Secondly, the widely used biomass burning method, BBM, was used to verify the results of the OCO-2 model and found that the biases were between 25.76% and 157.11%. The different results between BBM and the OCO-2 model were because of the biases of the smoke plumes area, smoke plume height, burned area, biomass fuel load, and even the different biological communities of the study region with the references.

Thirdly, based on the different CO_2 emission results, the advantages and limitations of the OCO-2 model in estimating CO_2 emissions from wildfires were analyzed and found that OCO-2 is a useful tool for fire-emission monitoring.

The present study proposed a new approach to estimate CO_2 emissions from wildfires using remote-sensing data and extends the application range of GHG satellites. In the future, many works are needed to perfect the OCO-2 model, especially to improve its efficiency and accuracy.

Author Contributions: Methodology, M.G.; software, L.W.; formal analysis, S.H.; writing—original draft preparation, M.G. and J.L.; writing—review and editing, M.G. and J.L.; funding acquisition, J.L., M.G.

Funding: This research was funded by the National Natural Science Foundation of China (41871103 and 41771179) and the National Key Research and Development Project (2016YFA0602301).

Acknowledgments: We would like to thank the NASA Goddard Earth Science Data and Information Services Center for the use of OCO-2 data in this study. We also would like to thank MISR data of NASA and the MINX v4.0 software of California Institute of Technology.

Conflicts of Interest: The authors declare that there is no conflict of interest.

References

1. Williams, C.A.; Gu, H.; MacLean, R.; Masek, J.G.; Collatz, G.J. Disturbance and the carbon balance of us forests: A quantitative review of impacts from harvests, fires, insects, and droughts. *Glob. Planet. Chang.* **2016**, *143*, 66–80. [CrossRef]

2. Boby, L.A.; Schuur, E.A.G.; Mack, M.C.; Verbyla, D.; Johnstone, J.F. Quantifying fire severity, carbon, and nitrogen emissions in Alaska's boreal forest. *Ecol. Appl.* **2010**, *20*, 1633–1647. [CrossRef] [PubMed]

3. Van Marle, M.J.E.; Kloster, S.; Magi, B.I.; Marlon, J.R.; Daniau, A.L.; Field, R.D.; Arneth, A.; Forrest, M.; Hantson, S.; Kehrwald, N.M.; et al. Historic global biomass burning emissions for CMIP6 (BB4CMIP) based on merging satellite observations with proxies and fire models (1750–2015). *Geosci. Model Dev.* **2017**, *10*, 3329–3357. [CrossRef]

4. Fisher, D.; Muller, J.P.; Yershov, V.N. Automated stereo retrieval of smoke plume injection heights and retrieval of smoke plume masks from aatsr and their assessment with CALIPSO and MISR. *IEEE Trans. Geosci. Remote Sens.* **2014**, *52*, 1249–1258. [CrossRef]

5. Langner, A.; Siegert, F. Spatiotemporal fire occurrence in Borneo over a period of 10 years. *Glob. Chang. Biol.* **2009**, *15*, 48–62. [CrossRef]

6. Alexander, H.D.; Mack, M.C. A canopy shift in interior Alaskan boreal forests: Consequences for above- and belowground carbon and nitrogen pools during post-fire succession. *Ecosystems* **2016**, *19*, 98–114. [CrossRef]

7. Guo, M.; Li, J.; Xu, J.; Wang, X.; He, H.; Wu, L. CO_2 emissions from the 2010 russian wildfires using gosat data. *Environ. Pollut.* **2017**, *226*, 60–68. [CrossRef] [PubMed]

8. Virkkula, A.; Levula, J.; Pohja, T.; Aalto, P.P.; Keronen, P.; Schobesberger, S.; Clements, C.B.; Pirjola, L.; Kieloaho, A.J.; Kulmala, L.; et al. Prescribed burning of logging slash in the boreal forest of Finland: Emissions and effects on meteorological quantities and soil properties. *Atmos. Chem. Phys.* **2014**, *14*, 4473–4502. [CrossRef]

9. Seiler, W.; Crutzen, P.J. Estimates of gross and net fluxes of carbon between the biosphere and the atmosphere from biomass burning. *Clim. Chang.* **1980**, *2*, 207–247. [CrossRef]

10. Miranda, A.I.; Coutinho, M.; Borrego, C. Forest fire emissions in Portugal: A contribution to global warming? *Environ. Pollut.* **1994**, *83*, 121–123. [CrossRef]

11. Wiedinmyer, C.; Neff, J.C. Estimates of CO_2 from fires in the United States: Implications for carbon management. *Carbon Balance Manag.* **2007**, *2*, 10. [CrossRef] [PubMed]

12. Barbosa, P.; Camia, A.; Kucera, J.; Libertà, G.; Palumbo, I.; San-Miguel-Ayanz, J.; Schmuck, G. *Chapter 8 Assessment of Forest Fire Impacts and Emissions in the European Union Based on the European Forest Fire Information System*; Elsevier Science & Technology: Amsterdam, The Netherlands, 2008; pp. 197–208.

13. Van der Werf, G.R.; Randerson, J.T.; Giglio, L.; Collatz, G.J.; Mu, M.; Kasibhatla, P.S.; Morton, D.C.; Defries, R.S.; Jin, Y.; Leeuwen, T.T.V. Global fire emissions and the contribution of deforestation, savanna, forest, agricultural, and peat fires (1997–2009). *Atmos. Chem. Phys.* **2010**, *10*, 16153–16230. [CrossRef]

14. Rosa, I.M.D.; Pereira, J.M.C.; Tarantola, S. Atmospheric emissions from vegetation fires in Portugal (1990–2008): Estimates, uncertainty analysis, and sensitivity analysis. *Atmos. Chem. Phys.* **2011**, *11*, 2625–2640. [CrossRef]

15. Shi, Y.; Yamaguchi, Y. A high-resolution and multi-year emissions inventory for biomass burning in Southeast Asia during 2001–2010. *Atmos. Environ.* **2014**, *98*, 8–16. [CrossRef]

16. Zhou, Y.; Xing, X.; Lang, J.; Chen, D.; Cheng, S.; Wei, L.; Wei, X.; Liu, C. A comprehensive biomass burning emission inventory with high spatial and temporal resolution in China. *Atmos. Chem. Phys.* **2017**, *17*, 2839–2864. [CrossRef]

17. Pereira, G.; Freitas, S.R.; Moraes, E.C.; Ferreira, N.J.; Shimabukuro, Y.E.; Rao, V.B.; Longo, K.M. Estimating trace gas and aerosol emissions over South America: Relationship between fire radiative energy released and aerosol optical depth observations. *Atmos. Environ.* **2009**, *43*, 6388–6397. [CrossRef]

18. Liu, M.; Song, Y.; Yao, H.; Kang, Y.; Li, M.; Huang, X.; Hu, M. Estimating emissions from agricultural fires in the North China Plain based on MODIS fire radiative power. *Atmos. Environ.* **2015**, *112*, 326–334. [CrossRef]

19. Wooster, M.J.; Freeborn, P.H.; Archibald, S.; Oppenheimer, C.; Roberts, G.J.; Smith, T.E.L.; Govender, N.; Burton, M.; Palumbo, I. Field determination of biomass burning emission ratios and factors via open-path ftir spectroscopy and fire radiative power assessment: Headfire, backfire and residual smouldering combustion in African savannahs. *Atmos. Chem. Phys.* **2011**, *11*, 11591–11615. [CrossRef]

20. Kaiser, J.W.; Heil, A.; Andreae, M.O.; Benedetti, A.; Chubarova, N.; Jones, L.; Morcrette, J.J.; Razinger, M.; Schultz, M.G.; Suttie, M. Biomass burning emissions estimated with a global fire assimilation system based on observed fire radiative power. *Biogeosci. Discuss.* **2012**, *9*, 527–554. [CrossRef]

21. Konovalov, I.B.; Berezin, E.V.; Ciais, P.; Broquet, G.; Beekmann, M.; Hadjilazaro, J.; Clerbaux, C.; Andreae, M.O.; Kaiser, J.W.; Schulze, E.D. Constraining CO_2 emissions from open biomass burning by satellite observations of co-emitted species: A method and its application to wildfires in Siberia. *Atmos. Chem. Phys.* **2014**, *14*, 10383–10410. [CrossRef]

22. Konovalov, I.B.; Beekmann, M.; Berezin, E.V.; Formenti, P.; Andreae, M.O. Probing into the aging dynamics of biomass burning aerosol by using satellite measurements of aerosol optical depth and carbon monoxide. *Atmos. Chem. Phys.* **2016**, *17*, 4513–4537. [CrossRef]

23. Mu, M.; Randerson, J.T.; Van der Werf, G.R.; Giglio, L.; Kasibhatla, P.; Morton, D.; Collatz, G.J.; Defries, R.S.; Hyer, E.J.; Prins, E.M. Daily and 3-hourly variability in global fire emissions and consequences for atmospheric model predictions of carbon monoxide. *J. Geophys. Res. Atmos.* **2011**, *116*, 24303. [CrossRef]

24. Wiedinmyer, C.; Akagi, S.K.; Yokelson, R.J.; Emmons, L.K. The Fire INventory from NCAR (FINN)—A high resolution global model to estimate the emissions from open burning. *Geosci. Model Dev. Discuss.* **2011**, *3*, 625–641. [CrossRef]

25. Kaiser, J.W.; Flemming, J.; Schultz, M.G.; Suttie, M.; Wooster, M.J. The MACC Global Fire Assimilation System: First emission products (GFASv0). *ECMWF Tech. Memo.* **2009**, *596*, 1–6.

26. Arnett, J.T.; Coops, N.C.; Daniels, L.D.; Falls, R.W. Detecting forest damage after a low-severity fire using remote sensing at multiple scales. *Int. J. Appl. Earth Obs. Geoinf.* **2015**, *35*, 239–246. [CrossRef]

27. Heymann, J.; Reuter, M.; Buchwitz, M.; Schneising, O.; Bovensmann, H.; Burrows, J.P.; Massart, S.; Kaiser, J.W.; Crisp, D. CO_2 emission of indonesian fires in 2015 estimated from satellite-derived atmospheric CO_2 concentrations. *Geophys. Res. Lett.* **2017**, *44*, 1537–1544. [CrossRef]

28. Patra, P.K.; Crisp, D.; Kaiser, J.W.; Wunch, D.; Saeki, T.; Ichii, K.; Sekiya, T.; Wennberg, P.O.; Feist, D.G.; Pollard, D.F. The orbiting carbon observatory (OCO-2) tracks 2–3 peta-gram increase in carbon release to the atmosphere during the 2014–2016 El Nino. *Sci. Rep.* **2017**, *7*, 13567. [CrossRef]

29. Connor, B.; Bösch, H.; McDuffie, J.; Taylor, T.; Fu, D.; Frankenberg, C.; O'Dell, C.; Payne, V.H.; Gunson, M.; Pollock, R.; et al. Quantification of uncertainties in OCO-2 measurements of XCO_2: Simulations and linear error analysis. *Atmos. Meas. Tech.* **2016**, *9*, 5227–5238. [CrossRef]

30. Wunch, D.; Wennberg, P.O.; Osterman, G.; Fisher, B.; Naylor, B.; Roehl, C.M.; O'Dell, C.; Mandrake, L.; Viatte, C.; Kiel, M. Comparisons of the Orbiting Carbon Observatory-2 (OCO-2) XCO_2 measurements with TCCON. *Atmos. Meas. Tech.* **2017**, *10*, 1–45. [CrossRef]

31. Crisp, D.; Pollock, H.; Rosenberg, R.; Chapsky, L.; Lee, R.; Oyafuso, F.; Frankenberg, C.; Dell, C.; Bruegge, C.; Doran, G.; et al. The on-orbit performance of the Orbiting Carbon Observatory-2 (OCO-2) instrument and its radiometrically calibrated products. *Atmos. Meas. Tech.* **2017**, *10*, 59–81. [CrossRef]

32. Pollock, R.; Haring, R.E.; Holden, J.R.; Johnson, D.L.; Kapitanoff, A.; Mohlman, D.; Phillips, C.; Randall, D.; Rechsteiner, D.; Rivera, J.; et al. The orbiting carbon observatory instrument: Performance of the oco instrument and plans for the OCO-2 instrument. In Proceedings of the SPIE—The International Society for Optical Engineering, Toulouse, France, 13 October 2010.

33. Prasad, P.; Rastogi, S.; Singh, R.P.; Panigrahy, S. Spectral modelling near the 1.6 μm window for satellite based estimation of CO_2. *Spectrochim. Acta A Mol. Biomol. Spectrosc.* **2014**, *117*, 330–339. [CrossRef] [PubMed]

34. Lee, R.A.M.; O'Dell, C.W.; Wunch, D.; Roehl, C.M.; Osterman, G.B.; Blavier, J.F.; Rosenberg, R.; Chapsky, L.; Frankenberg, C.; Hunyadi-Lay, S.L.; et al. Preflight spectral calibration of the Orbiting Carbon Observatory 2. *IEEE Trans. Geosci. Remote Sens.* **2017**, *55*, 2499–2508. [CrossRef]

35. Day, J.O.; O'Dell, C.W.; Pollock, R.; Bruegge, C.J.; Rider, D.; Crisp, D.; Miller, C.E. Preflight spectral calibration of the orbiting carbon observatory. In Proceedings of the International Conference on Next Generation Mobile Applications, Amman, Jordan, 27–29 July 2010; pp. 13–18.

36. Frankenberg, C.; Pollock, R.; Lee, R.A.M.; Rosenberg, R.; Blavier, J.F.; Crisp, D.; O'Dell, C.W.; Osterman, G.B.; Roehl, C.; Wennberg, P.O.; et al. The Orbiting Carbon Observatory (OCO-2): Spectrometer performance evaluation using pre-launch direct sun measurements. *Atmos. Meas. Tech.* **2015**, *8*, 301–313. [CrossRef]

37. Meng, J.; Ding, G.; Liu, L.; Zhang, R. A comparison and validation of atmosphere CO_2 concentration OCO-2-based observations and tccon-based observations. In *Communications in Computer and Information Science*; Springer: Singapore, 2016; Volume 645, pp. 356–363.

38. Eldering, A.; O'Dell, C.W.; Wennberg, P.O.; Crisp, D.; Gunson, M.R.; Viatte, C.; Avis, C.; Braverman, A.; Castano, R.; Chang, A.; et al. The Orbiting Carbon Observatory-2: First 18 months of science data products. *Atmos. Meas. Tech.* **2017**, *10*, 549–563. [CrossRef]

39. Mandrake, L.; Frankenberg, C.; O'Dell, C.W.; Osterman, G.; Wennberg, P.; Wunch, D. Semi-autonomous sounding selection for OCO-2. *Atmos. Meas. Tech.* **2013**, *6*, 2851–2864. [CrossRef]

40. Ross, A. *Gosat Measurements of Wildfire Emissions*; University of London: London, UK, 2012.

41. Ross, A.N.; Wooster, M.J.; Boesch, H.; Parker, R. First satellite measurements of carbon dioxide and methane emission ratios in wildfire plumes. *Geophys. Res. Lett.* **2013**, *40*, 4098–4102. [CrossRef]

42. Chand, T.R.K.; Badarinath, K.V.S.; Murthy, M.S.R.; Rajshekhar, G.; Elvidge, C.D.; Tuttle, B.T. Active forest fire monitoring in Uttaranchal State, India using multi-temporal DMSP-OLS and MODIS data. *Int. J. Remote Sens.* **2007**, *28*, 2123–2132. [CrossRef]

43. Guo, M.; Li, J.; Sheng, C.; Xu, J.; Li, W. A review of wetland remote sensing. *Sensors* **2017**, *17*, 777. [CrossRef]

44. Roy, D.; Descloitres, J.; Alleaume, S. The MODIS fire products. *Remote Sens. Environ.* **2002**, *83*, 244–262.

45. Cheng, D.; Rogan, J.; Schneider, L.; Cochrane, M. Evaluating MODIS active fire products in subtropical yucatán forest. *Remote Sens. Lett.* **2013**, *4*, 455–464. [CrossRef]

46. Chen, D.; Pereira, J.M.C.; Masiero, A.; Pirotti, F. Mapping fire regimes in China using MODIS active fire and burned area data. *Appl. Geogr.* **2017**, *85*, 14–26. [CrossRef]

47. Ganci, G.; Negro, C.D.; Fortuna, L.; Vicari, A. A tool for multi-platform remote sensing processing. *Commun. Simai Congr.* **2009**, *3*, 281.

48. Wan, Z.; Ng, D.; Dozier, J. Spectral emissivity measurements of land-surface materials and related radiative transfer simulations. *Adv. Space Res.* **1994**, *14*, 91–94. [CrossRef]

49. Mazzoni, D.; Logan, J.A.; Diner, D.; Kahn, R.; Tong, L.; Li, Q. A data-mining approach to associating MISR smoke plume heights with MODIS fire measurements. *Remote Sens. Environ.* **2007**, *107*, 138–148. [CrossRef]

50. Kahn, R.A.; Moroney, C.M.; Gaitley, B.J.; Nelson, D.L.; Garay, M.J.; Mims, S.R. MISR stereo heights of grassland fire smoke plumes in Australia. *IEEE Trans. Geosci. Remote Sens.* **2009**, *48*, 25–35.

51. Gonzalez, L.; Val Martin, M.; Kahn, R. Biomass burning smoke heights over the Amazon observed from space. *Atmos. Chem. Phys.* **2019**, *19*, 1685–16702. [CrossRef]

52. Nelson, D.L.; Garay, M.J.; Kahn, R.A.; Dunst, B.A. Stereoscopic height and wind retrievals for aerosol plumes with the MISR INteractive eXplorer (MINX). *Remote Sens.* **2013**, *5*, 4593–4628. [CrossRef]

53. Sofiev, M.; Ermakova, T.; Vankevich, R. Evaluation of the smoke-injection height from wild-land fires using remote-sensing data. *Atmos. Chem. Phys.* **2012**, *11*, 27937–27966. [CrossRef]

54. Kahn, R.A.; Li, W.H.; Moroney, C.; Diner, D.J.; Martonchik, J.V.; Fishbein, E. Aerosol source plume physical characteristics from space-based multiangle imaging. *J. Geophys. Res. Atmos.* **2007**, *112*, 1–20.

55. Goto, Y.; Suzuki, S. Estimates of carbon emissions from forest fires in Japan, 1979–2008. *Int. J. Wildland Fire* **2013**, *22*, 721–729. [CrossRef]

56. Chang, D.; Song, Y. Estimates of biomass burning emissions in tropical Asia based on satellite-derived data. *Atmos. Chem. Phys.* **2010**, *10*, 2335–2351. [CrossRef]

57. Shi, Y.; Matsunaga, T.; Yamaguchi, Y. High-resolution mapping of biomass burning emissions in three tropical regions. *Environ. Sci. Technol.* **2015**, *49*, 10806–10814. [CrossRef] [PubMed]

58. Shi, Y.; Sasai, T.; Yamaguchi, Y. Spatio-temporal evaluation of carbon emissions from biomass burning in Southeast Asia during the period 2001–2010. *Ecol. Model.* **2014**, *272*, 98–115. [CrossRef]

59. Akagi, S.K.; Yokelson, R.J.; Wiedinmyer, C.; Alvarado, M.J.; Reid, J.S.; Karl, T.; Crounse, J.D.; Wennberg, P.O. Emission factors for open and domestic biomass burning for use in atmospheric models. *Atmos. Chem. Phys.* **2011**, *11*, 4039–4072. [CrossRef]

60. Vadrevu, K.; Ellicott, E.; Giglio, L.; Badarinath, K.V.S.; Vermote, E.; Justice, C.; Lau, W. Vegetation fires in the himalayan region—aerosol load, black carbon emissions and smoke plume heights. *Atmos. Environ.* **2012**, *47*, 241–251. [CrossRef]

61. Wu, Y.; Cordero, L.; Gross, B.; Moshary, F.; Ahmed, S. Smoke plume optical properties and transport observed by a multi-wavelength lidar, sunphotometer and satellite. *Atmos. Environ.* **2012**, *63*, 32–42. [CrossRef]

62. Saatchi, S.S.; Harris, N.L.; Brown, S.; Lefsky, M.; Mitchard, E.T.; Salas, W.; Zutta, B.R.; Buermann, W.; Lewis, S.L.; Hagen, S. Benchmark map of forest carbon stocks in tropical regions across three continents. *Proc. Natl. Acad. Sci. USA* **2011**, *108*, 9899. [CrossRef] [PubMed]

63. Kauffman, J.B.; Steele, M.D.; Cummings, D.L.; Jaramillo, V.J. Biomass dynamics associated with deforestation, fire, and conversion to cattle pasture in a Mexican tropical dry forest. *For. Ecol. Manag.* **2003**, *176*, 1–12. [CrossRef]

64. O'Dell, C.W.; Connor, B.; Bösch, H.; O'Brien, D. The ACOS CO_2 retrieval algorithm—Part 1: Description and validation against synthetic observations. *Atmos. Meas. Tech.* **2011**, *4*, 99–121.

65. Yurganov, L.N.; Rakitin, V.; Dzhola, A.; August, T.; Fokeeva, E.; George, M.; Gorchakov, G.; Grechko, E.; Hannon, S.; Karpov, A. Satellite- and ground-based CO total column observations over 2010 Russian fires: Accuracy of top-down estimates based on thermal IR satellite data. *Atmos. Chem. Phys. Discuss.* **2011**, *11*, 7925–7942. [CrossRef]

Article

Evaluation of Different WRF Parametrizations over the Region of Iași with Remote Sensing Techniques

Iulian-Alin Roșu [1], Silvia Ferrarese [2], Irina Radinschi [3], Vasilica Ciocan [4] and Marius-Mihai Cazacu [3,*]

[1] Faculty of Physics, "Alexandru Ioan Cuza" University of Iasi, Bulevardul Carol I 11, 700506 Iasi, Romania; alin.iulian.rosu@gmail.com
[2] Department of Physics, University of Turin, via P. Giuria 1, 10125 Turin, Italy; silvia.ferrarese@unito.it
[3] Department of Physics, "Gheorghe Asachi" Technical University of Iasi, 700050 Iasi, Romania; radinschi@yahoo.com
[4] Faculty of Civil Engineering and Buiding Services, "Gheorghe Asachi" Technical University of Iasi, 700050 Iasi, Romania; vciocan2005@yahoo.com
* Correspondence: cazacumarius@gmail.com

Received: 26 July 2019; Accepted: 16 September 2019; Published: 18 September 2019

Abstract: This article aims to present an evaluation of the Weather Research and Forecasting (WRF) model with multiple instruments when applied to a humid continental region, in this case, the region around the city of Iași, Romania. A series of output parameters are compared with observed data, obtained on-site, with a focus on the Planetary Boundary Layer Height (PBLH) and on PBLH-related parametrizations used by the WRF model. The impact of each different parametrization on physical quantities is highlighted during the two chosen measurement intervals, both of them in the warm season of 2016 and 2017, respectively. The instruments used to obtain real data to compare to the WRF simulations are: a lidar platform, a photometer, and ground-level (GL) meteorological instrumentation for the measurement of temperature, average wind speed, and pressure. Maps of PBLH and 2 m above ground-level (AGL) atmospheric temperature are also presented, compared to a topological and relief map of the inner nest of the WRF simulation. Finally, a comprehensive simulation performance evaluation of PBLH, temperature, wind speed, and pressure at the surface and total precipitable water vapor is performed.

Keywords: WRF; PBL; lidar; photometry; simulation; meteorology

---

1. Introduction

Simulating atmospheric conditions through meteorological modeling presents multiple advantages in the domain of weather forecasting, such as near-unlimited geographical versatility and applicability; however, the mathematical complexity associated with these models and the quasi-chaotic nature of the atmosphere guarantees that such calculations will always contain a degree of uncertainty [1,2]. Modeling meteorological quantities in the Planetary Boundary Layer (PBL) is even more difficult but also crucial for air quality analysis and forecast. In fact, the thermodynamic state of the PBL plays a significant role in mixing and dispersing of air pollutant. Temperature, wind speed, precipitable water vapor, and PBL height (PBLH) are key-quantities describing the daily evolution of PBL structure. A potential method for improving meteorological modeling consists of correlating atmospheric simulation and observation, in which case, we find telemetry and other non-intrusive techniques advantageous, along with more traditional techniques [3–5].

Many recent studies investigated the role of PBLH parametrizations in simulations. Several authors studied the sensitivity of the PBL schemes available in the Weather Research and Forecasting (WRF) model in simulating the daily evolution of PBL [4,6–10]. The direct comparison of simulated temperature and humidity data against observed data is generally performed and discussed alongside the comparison between simulated and measured PBLH values, where PBLH measured data are available [4,6,9]. Otherwise, if PBLH observations were not available, the intercomparison between values from different numerical experiments provided the relative performance of different PBL schemes [8,10]. Note that PBLH measurements can be carried out with lidar systems working at high temporal frequency and vertical spatial coverage [7].

Numerical weather prediction models work poorly over complex terrain where the exchange in energy and mass is not restricted to vertical turbulent mixing as over flat, homogenous and horizontal terrain. Many previous studies evaluated the performance of model PBL parametrization schemes in locations known for complex atmospheric situations [11,12]. In one study, the influence of three PBL schemes from the legacy Fifth Generation Penn State-NCAR Mesoscale Model on meteorological and air quality simulations over Barcelona was analysed [13]. The authors found that the MM5 model tended to show a cold bias, with higher model-simulated wind speeds compared with observations, depending on the PBL scheme used. Other studies were focused on the influence of five PBL parametrizations on air-quality predictions over the greater Athens area [12], and on the evaluation of WRF model-simulated PBLH over Barcelona using eight PBL schemes [7]. Model-simulated PBLH was validated with PBLH estimates from a backscatter lidar during a 7-year period. The authors determined that a non-local scheme such as the Asymmetrical Convective Model version 2 (ACM2) provides the most accurate simulations of PBLH, even under diverse synoptic flows such as regional recirculation.

In the present work, the investigated area is located in Romania (Iași province). The area is characterized by complex terrain containing cities, forest, and grassland and crossed by rivers. Due to the presence of multiple hills and riverbeds the altitude ranges from about 1 m to 600 m a.s.l. Several WRF studies have taken place over specific regions of Romania, or over the territory of Romania and the nearby countries in general. In particular, a similar study over the Southern Carpathians had the objective of evaluating WRF GL or near-surface output [14]; it was found that throughout much of the WRF simulation, average wind speed was severely overestimated, and error minimization can be achieved by properly selecting physical configurations suitable for the region at hand [14]. A study is concerned with the analysis of the urban heat island of Bucharest present in WRF output temperature maps [15], and another compares WRF solar irradiation in Romania with observed values from different stations over a determined period [16]. As a final example, a study targeting the capital of the neighboring country of Bulgaria attempts to use high-resolution WRF simulations to quantify the impact of urbanization on local meteorological conditions, showing, in particular, a significant increase of temperature in the central Sofia area [17].

A total of twelve WRF simulations were performed for two episodes and evaluated with lidar, photometry, and ground-level meteorological data. For every episode, six simulations, each with a different set of parametrizations linked to the nature and behaviour of the PBL, were carried out [18]. The model outputs used in this analysis were PBL height, total precipitable water, GL temperature, GL pressure, and GL average wind speed. Lidar data was used to extract the PBLH, sun-photometer data was used to obtain real total precipitable water, and standard meteorological instrumentation was used to obtain observed GL values [19,20].

The sets of observed and simulated data were compared by calculating the mean absolute difference, mean bias error, mean square error and the correlation coefficient between them. Error calculus and data plotting were carried out with software developed in Python 3.5.

2. WRF Model Setup and Target Area

The WRF model is a next-generation mesoscale numerical weather prediction system designed for both atmospheric research and operational forecasting applications [21,22]. The model serves a wide range of meteorological applications, featuring two dynamical cores: a data assimilation system, and a software architecture supporting parallel computation and system extensibility [21].

In this study, we used WRF version 3.9.1 with the Advanced Research WRF (ARW) dynamical solver [23]. Initial and boundary conditions were provided by IFS (Integrating Forecasting System) 6-hours analyses from ECMWF (European Centre for Medium-Range Weather Forecasts) with horizontal grid spacing of 0.125° × 0.125° in latitude and longitude. Three nested model domains were configured with varying horizontal grid spacing at the parent European level (9 km × 9 km; 120 × 120 grid points), and two nested domains roughly encompassing the Romania-Moldova region (3 km × 3 km; 106 × 106 grid points) and the Iași county, along with the Moldavian Ungheni district (1 km × 1 km; 94 × 94 grid points) (Figures 1 and 2). It is assumed that 1 km × 1 km grid spacing is of fine enough detail to resolve most mesoscale features in the complex study area [24]. In the simulations, 50 vertical levels were arranged between the surface and the 100 hPa upper boundary.

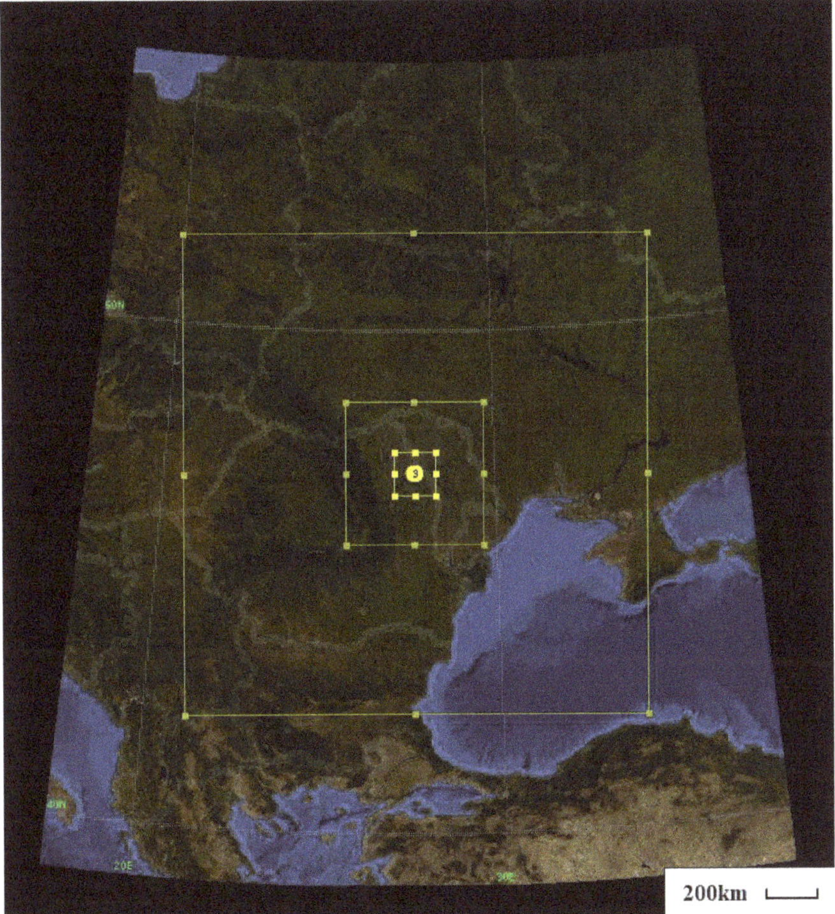

Figure 1. Three nested model domains used in this study (yellow lines). 1st nest: contains the entirety of Romania and Republic of Moldova, along with the area around it and sections of neighboring countries. 2nd nest: contains larger Moldova and Bucovina region of Romania, and the majority of the Republic of Moldova. 3rd nest: contains area around the city of Iași. Number 3 (in yellow circle) represents the city of Iași.

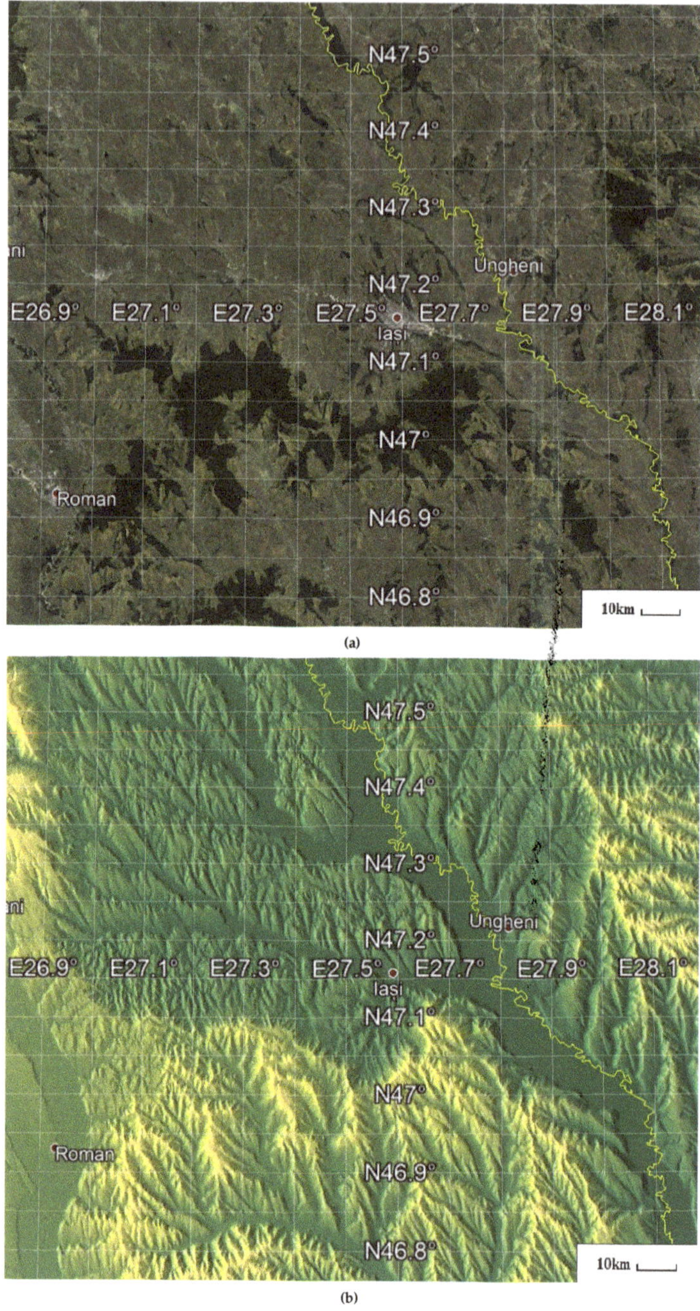

Figure 2. (**a**) Satellite map of the 3rd nest with latitude/longitude grid; border between Romania (right) and Moldova (left) delineated by the Prut river (yellow line); Google Earth. (**b**) Relief map of the 3rd nest with latitude/longitude grid; border between Romania (right) and Moldova (left) delineated by the Prut river (yellow line); altitude colormap: ~5 m above ground-level (AGL) dark green, ≤300 m AGL bright yellow; Google Earth.

Two events were studied, the first occurring from 16 May 2017 at 18:00 UTC to 17 May 2017 at 12:20 UTC and the second on the 4 April 2016 from 00:00 to 18:20 UTC. These two events were characterised by clear sky and a lack of precipitation; registered temperature, pressure and average wind speed show a relatively calm atmosphere. Atmospheric conditions are mostly fair for both events. Given that all the real data used to compare simulated WRF data are collected from instruments located in the city of Iași, data from the centre of the 3rd nest and the particularities of this nest will form the main objective of analysis in this study.

The geographical and seasonal particularities of the chosen area provide an interesting context and justification for this study. The area delineated by the 3rd nest is, under the Köppen–Geiger climate classification system [25], classified as Dfb (humid continental, without dry season, with warm summer) [26]. Multiple hills dot the area, with the highest point in the nest being at 605 m AGL, at Hîrtop Hill, and the lowest point being 2 m AGL, near the village of Gorban. Most of the rivers and rivulets (such as Bahlui and Jijia) flow south-eastwards into the larger Prut river, the natural border between Romania and the Republic of Moldova. The southern part of the nested area is heavily forested, while the northern part contains a large number of small bodies of water, such as the lakes Bulbucani, Hălcești, and even a fish farm destined for pisciculture and research [27]. In any case, the majority of the area is complex continental landmass, which can pose a challenge for the model; as opposed to simulations above flat, horizontal and homogeneous terrain, where the PBL height varies relatively slowly in space [28,29]. In terms of anthropogenic activity, the majority of the area is rural, with the largest by far urban area being the city of Iași, positioned at the centre of the nested area. The city has the second largest population in Romania [30], and a booming industrial and infrastructure sector [31]. According to the 2018 AirVisual World Air Quality Report, Iași is the most polluted city in Romania [32]. Topologically, the city stretches over at least seven hills, and is traversed by the river Bahlui. Overall, the anthropogenic influence of the central city and the geographic diversity of the area make for a suitable and interesting target for WRF simulations; especially in terms of the evolution of the PBL height in time.

In the present work, the sensitivity of some PBL schemes has been analysed. The PBL parametrizations available in WRF model and applied here consist of: YSU (YonSei University) scheme [33], MYJ (Mellor–Yamada–Janjic) scheme [34], ACM2 (Asymmetrical Convective Model version 2) scheme [35], BouLac (Bougeault–Lacarrère) scheme [36], TEMF (Total Energy–Mass Flux) scheme [37] and ShinHong scheme [38]. The first and most widely used PBL scheme is the YSU. It is a first-order, non-local scheme with an explicit entrainment layer and a parabolic K-profile in an unstable mixed layer, where PBLH in the YSU scheme is determined from the *Rib* (bulk Richardson number) method but calculated starting from the surface. The MYJ scheme is a one-and-a-half order prognostic TKE scheme with local vertical mixing. The ACM2 scheme is a first-order, non-local closure scheme and features non-local upward mixing and local downward mixing. This scheme has an eddy-diffusion component in addition to the explicit non-local transport of ACM1 (Asymmetrical Convective Model version 1). PBLH is determined as the height where the Rib calculated above the level of neutral buoyancy exceeds a critical value (Ribc = 0.25). For stable or neutral flows the scheme shuts off non-local transport and uses local closure. The BouLac scheme is a one-and-a-half order, local closure scheme and has a TKE prediction option designed for use with the BEP (Building Environment Parametrization) multi-layer, urban canopy model [39]. BouLac diagnoses PBLH as the height where the prognostic TKE reaches a sufficiently small value (in the current version of WRF is 0.005 m^2s^{-2}). The TEMF scheme is a one-and-a-half order, non-local closure scheme and has a sub-gridscale total energy prognostic variable, in addition to mass-flux-type shallow convection. TEMF uses eddy diffusivity and mass flux concepts to determine vertical mixing. PBLH is calculated through a *Rib* method with zero as a threshold value. Finally, the Shin-Hong PBL scheme is a scale-aware scheme. The subgrid scale transport profile is parameterized based on the 2013 conceptual derivation documented in Shin and Hong [38]. First, the nonlocal transport by strong updrafts and local transport by the remaining small-scale eddies are separately calculated. Second, the subgrid scale nonlocal transport is formulated by multiplying a grid-size dependency function with the total nonlocal transport profile fitted to the large eddy simulation output.

Other parametrizations used in this study are as follows: short and long-wave radiation parametrizations are New Goddard Shortwave and Longwave Schemes. No cumulus parametrization is implemented. The microphysics parametrization is represented by the Milbrandt–Yau Double Moment Scheme, and the land surface model used is the Unified Noah Land Surface Model. Chosen feedback is 1; this results in two-way nesting.

3. Instrumentation

A variety of measuring equipment is used in this study to contrast and compare the influences of different parametrizations to the WRF output; the first of them is a lidar. These systems are utilized in the fields of high-resolution mapping, geodesy, archaeology, geography, atmospheric physics and many more. Performing atmospheric altitude profiles of meteorological parameters in general demands specialized instrumentation and data processing. A standard and reliable method to directly obtain meteorological parameters throughout the free atmosphere for low to high altitudes is by launching specialized weather balloons equipped with in-situ sensors [40]. Radar, sodar, and microwave radiometer platforms have also been used to determine PBLH, or other boundary layer-related quantities, in other studies [41–43]. Lidar systems combine many of the advantages gained from using these instruments into one remote sensing platform, that is capable of returning atmospheric data profiles with a high temporal and spatial resolution [44–46].

The technical specifications of the main components of the lidar platform utilized in the study are as follows: the laser component is a Nd:YAG, producing pulses of laser at a frequency of 30 Hz, with a wavelength of 532 nm, laser beam diameter of 6 mm and a pulse energy of 100 mJ; meanwhile, the optical component is a Newtonian LightBridge telescope with a primary mirror diameter of 406 mm. The signal overlap altitude in most cases is approximately 200 m, much lower than most PBLH instances in general, and lower than all PBLH instances measured in this study. The lidar platform utilized in the study has a spatial resolution of 3.75 m and both technical details and results from previous measurement campaigns were reported in the scientific literature [47,48].

A common use for a lidar system, in the context of atmospheric physics, is the obtaining of PBLH data [49,50]; in this study, a method detailed in one of our previous studies is used [44]. This method consists of producing an arbitrary equation that highlights the PBL as a clear and sharp peak, which allows us to extract its height and plot its evolution in time [44]:

$$BL(z) = \sigma_I^2(z)\frac{d\beta(z)}{dz}$$

in which $\sigma_I^2(z)$ is the scintillation profile of the RCS (Range Corrected Signal) obtained by lidar, and $\beta(z)$ is the backscatter coefficient profile obtained from RCS data via the Fernald-Klett inversion [51]. Our previous investigation shows that this equation can be used to retrieve the PBL height with greater confidence than other established methods, such as the gradient and the variance methods [44,49,50]. Regarding the potential influence of noise-related errors to the calculation of the scintillation profile, the overall signal uncertainty added by noise is:

$$\Delta V = \sqrt{NSF^2\left(V - \overline{V_b}\right) + (\Delta V_b)^2}$$

where V is raw lidar signal, V_b is background lidar signal, and NSF is the "noise scale factor", which is equal to the standard deviation of the shot noise divided by the square root of average shot noise [52]. It is determined that the signal uncertainty, in the case of attenuated backscatter, has values of the order 10^{-7} or lower [52], while a typical attenuated backscatter profile has values of the order 10^{-5} or lower; thus, this uncertainty represents variations hundreds of times smaller than the actual values of the profile. Since the model subtracts the "dark" signal (generated by photocathode thermionic emission, which is collected before the measurements begin [53]) from the raw signal, we can assume that such uncertainty is even lower. Also, the fact that the photomultiplier component of the lidar platform utilized in this study was being operated in analogue mode removes the need to consider possible instances of "afterpulsing", which only takes place when a photomultiplier component is operated in a "pulse detection mode" [53]. Finally, the fact that the photomultiplier component of the lidar platform used in this study is a PMT (photomultiplier tube) presents an advantage, since excess noise decreases with an increase in the average photo-multiplication gain in PMTs [52].

Another instrument that was used for data collecting and comparing purposes in this study is the sunphotometer, which is a type of photometer conceived in such a way that it points at the sun; the instrument used in the study is a Cimel Automatic Sun Tracking Photometer CE 318, which is a part of our AERONET Iasi_LOASL site [54,55]. While many of the higher applications of this instrument revolve around exploring aerosol properties in a given time and region [56,57], in this application the desired output of the instrument is the total precipitable water. Lidar and sunphotometer systems are sometimes used in conjunction [58]; a particular use for this coupling is to calculate the atmospheric aerosol backscatter profile [44,58]. The other instruments used to obtain GL data are located in the "Podul de Piatra" sector of Iași; data has been obtained from the ANM (Romanian National Meteorological Administration). The lidar, the photometer, and the GL instruments are located in three different sites, all of them situated, at most, 3 grid squares away from the center of the 3rd nest. All the measured values are compared accordingly to the simulated values in their own grid squares. There is just one GL monitoring station which contains all of the GL instruments, and there are no nearby natural or man-made structures that could present any obstruction to the instruments.

All operations of data manipulation and correlation were performed in Python 3.5, with Savitsky–Golay processing of real data profiles included in the plotting; this processing is a digital filter which can be applied to a set of points for the purpose of smoothing the data, through convolution [59,60]. The filter window chosen for this analysis is 17, and the order of the calculated fitting polynomial is 3; these numbers were chosen arbitrarily, in order to best fit the raw data from a graphical perspective. Finally, using the netCDF4 library, data from each WRF output file was extracted and compared to real data via plotting and evaluating statistical parameters.

4. Methods

In this section, the simulated data and instrument-related data are compared. The parameters of interest, extracted from WRF output, are the planetary boundary layer height (PBLH), atmospheric temperature at 2 m AGL (T2M), average wind speed at 10 m AGL (U10M), atmospheric pressure at 2 m AGL (P2M), and the total precipitable water vapor (WV). This last value refers to the total atmospheric water vapor contained in a vertical column of unit cross-sectional area extending between any two specified levels, commonly expressed in terms of the height to which that water substance would stand if completely condensed and collected [61]. Following this definition, the total precipitable water vapor can be generally calculated with:

$$WV = \frac{1}{\rho_{H2O}\,g} \int_{p1}^{p2} x(p)\,dp$$

where ρ_{H2O} is the density of water, g gravitational acceleration, and $x(p)$ the mixing ratio at a pressure level p [61]. This definition is used since the mixing ratio profile is given as straight-forward output by the WRF simulation. Considering $p1$ as the GL pressure, and $p2$ the pressure at the maximum altitude at which WRF simulates the mixing ratio, the previous equation is re-written as:

$$WV = 1.0228{\cdot}10^{-4} \int_{p_{GL}}^{p_{max}} x(p)\,dp$$

Note that the use of this metric can imply the assumption that all the water vapor in the analyzed region is concentrated in the PBL; it is certain that if this were the case, PBLH parametrizations would play a much higher role in influencing simulated WV. However, while this assumption is not made here, it is fair to assume that any small change in the water vapor content of the atmosphere would influence the PBL in some manner, and the opposite can also be assumed. Indeed, it can be seen in the following results segment that changes in PBLH parametrizations slightly influence the simulated WV. The statistical measures used to verify and contrast the WRF simulations are the mean absolute error (MAE), mean bias error (MBE), root mean square error (RMSE) and coefficient of determination (R^2).

5. Results

PBLH and T2M maps in the inner domain were obtained from the WRF simulations as examples. The outputs computed with YSU parametrization at 08:00 UTC in both episodes are shown in Figures 3–6. The temperature at 2 meters from the surface is lower and less variable on 17 May 2017 (Figure 5) than on 4 April 2016 (Figure 6) when the maximum values are reached along and across the Prut river basin, where the altitude is lowest. As a consequence, the PBLH maps tend to show higher values in regions where their respective T2M maps show higher temperature; for the 2017 scenario, that is roughly the region of the Prut river basin, and for the 2016 scenario, the hills south of Iași.

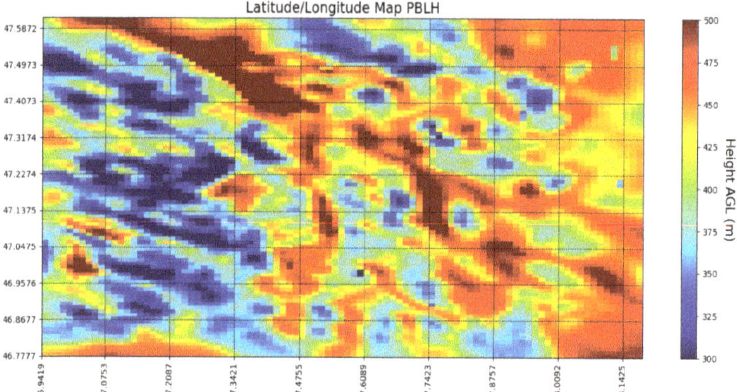

Figure 3. Example of a Weather Research and Forecasting (WRF) output map (PBLH); 3rd nest, in center of nest: Iași; 17/05/2017, 08:00 UTC; YSU parametrization.

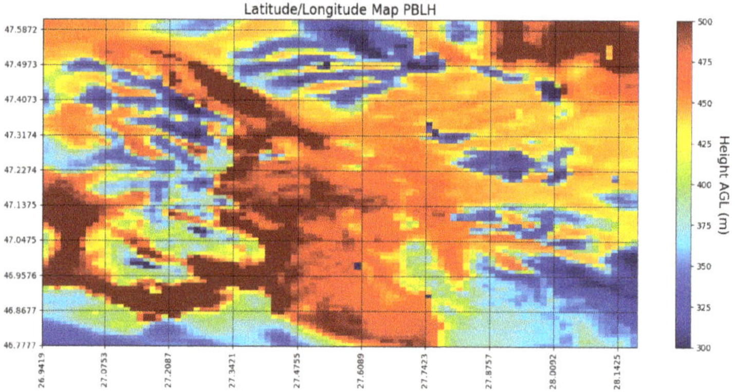

Figure 4. Example of a WRF output map (PBLH); 3rd nest, in center of nest: Iași; 04/04/2016, 08:00 UTC; YSU parametrization.

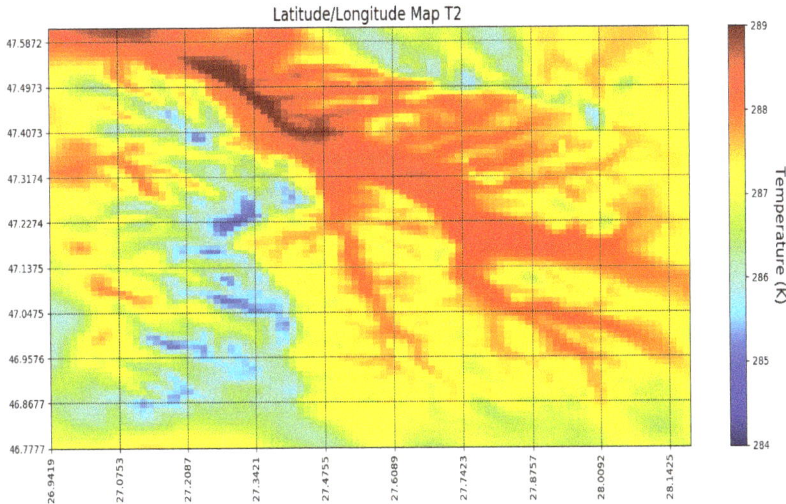

Figure 5. Example of a WRF output map (temperature at 2 m AGL); 3rd nest, in center of nest: Iași; 17/05/2017, 08:00 UTC; YSU parametrization.

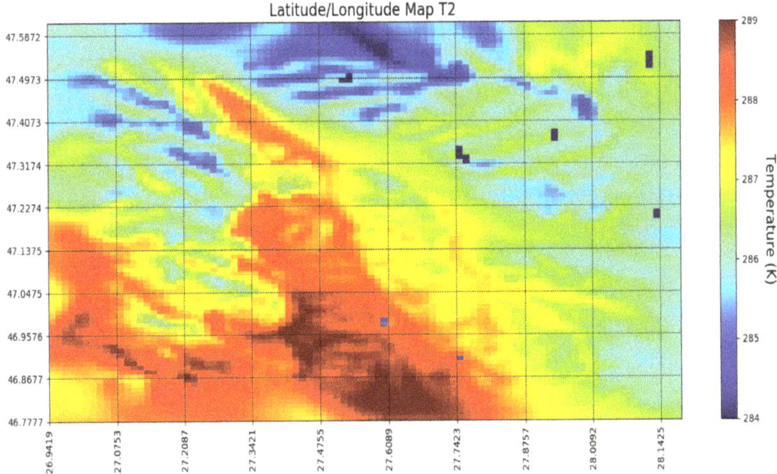

Figure 6. Example of a WRF output map (temperature at 2 m AGL); 3rd nest, in center of nest: Iași; 04/04/2016, 08:00 UTC; YSU parametrization.

In the following, the WRF output and the obtained data comparison are introduced by means of both graphical and table data. First, two examples of RCS data for each scenario are presented, in order to exemplify the typical signal obtained by the platform used in this study, and to determine that the PBLH is not lower than the complete signal overlap altitude (Figure 7). WRF results on 17 May 2017 show that the different parametrizations compute a nocturnal PBLH lower than 500 m and a diurnal that is always higher than the observed one (Figure 8). During the 4 April 2016 episode simulated PBLH values are lower than 200 m and their maxima occur between 13:00 and 15:00 UTC.

The comparison with observation shows that simulated PBLH values are higher than observed ones in the morning, and in agreement with observations in the afternoon (Figure 9). The PBL parametrizations have similar performances with the exception of TEMF, during 4 April 2016 episode, which seems to not catch the general PBLH trend. MAE and RMSE values between each parametrization and lidar data, and each parametrization with each other (Tables 1 and 2) indicate that MYJ is the best performing parametrization in the episode of 17 May 2017; the same is true for YSU and BouLac parametrizations in the episode of 4 April 2016, while R^2 values show a generally high correlation, with ACM2 performing best for the 2017 episode, and ShinHong performing best for the 2016 episode.

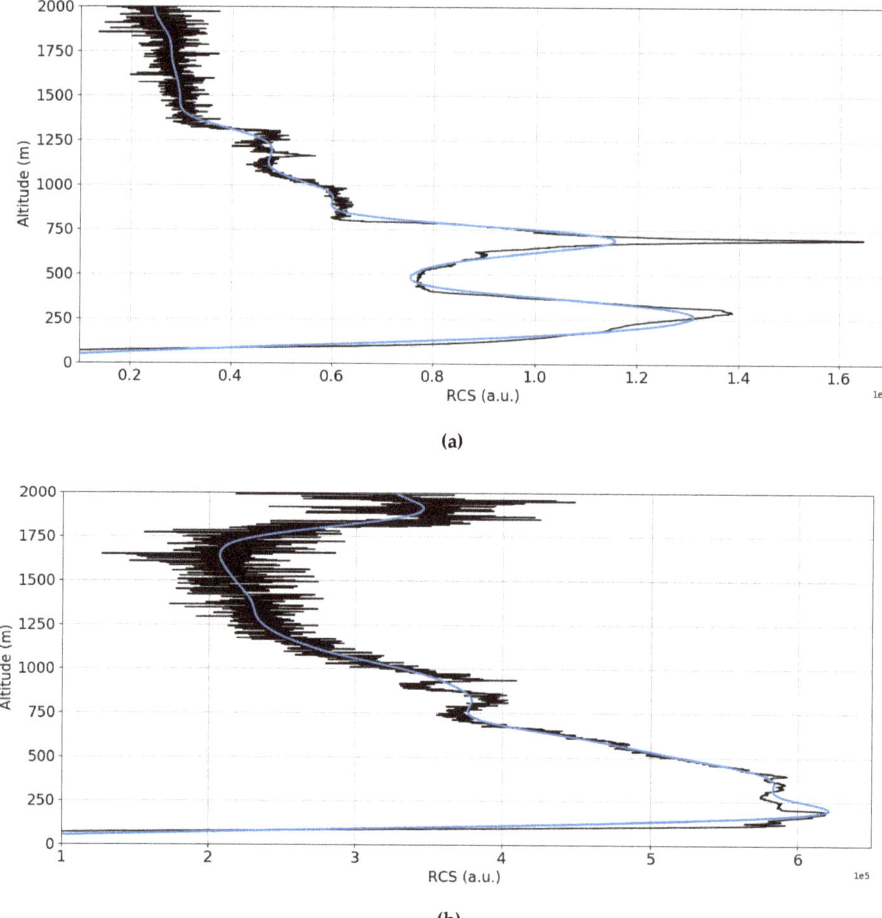

Figure 7. (**a**) Lidar-obtained RCS profile, 17/05/2017, 08:00 UTC, black: real data, blue: Savitsky–Golay smoothing of real data. (**b**) Lidar-obtained RCS profile, 04/04/2016, 08:00 UTC, black: real data, blue: Savitsky–Golay smoothing of real data.

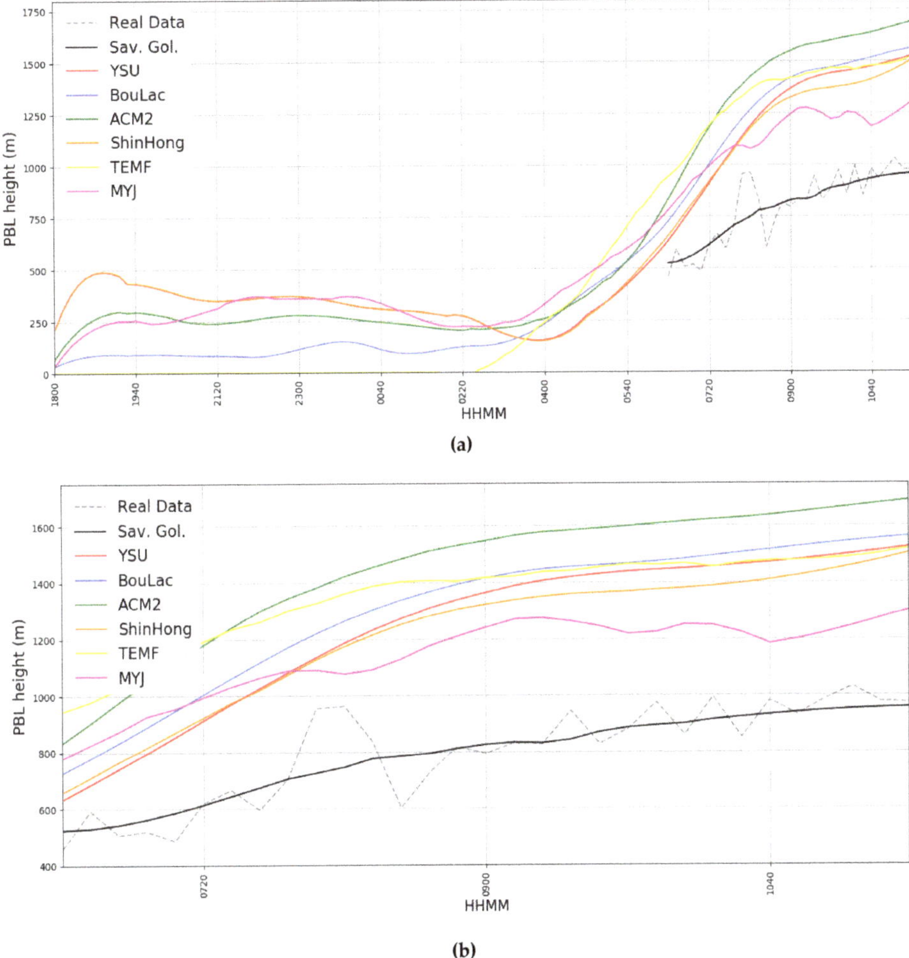

(a)

(b)

Figure 8. (**a**) Simulated PBLH timeseries compared with lidar-retrieved PBLH timeseries, 17/05/2017, dotted grey: real data, black: Savitsky–Golay smoothing of real data, red: simulated data with YSU parametrization, blue: simulated data with BouLac parametrization, green: simulated data with ACM2 parametrization, orange: simulated data with ShinHong parametrization, yellow: simulated data with TEMF parametrization, purple: simulated data with MYJ parametrization. (**b**) Simulated PBLH timeseries compared with lidar-retrieved PBLH timeseries, zoomed-in, 17/05/2017, dotted grey: real data, black: Savitsky–Golay smoothing of real data, red: simulated data with YSU parametrization, blue: simulated data with BouLac parametrization, green: simulated data with ACM2 parametrization, orange: simulated data with ShinHong parametrization, yellow: simulated data with TEMF parametrization, purple: simulated data with MYJ parametrization.

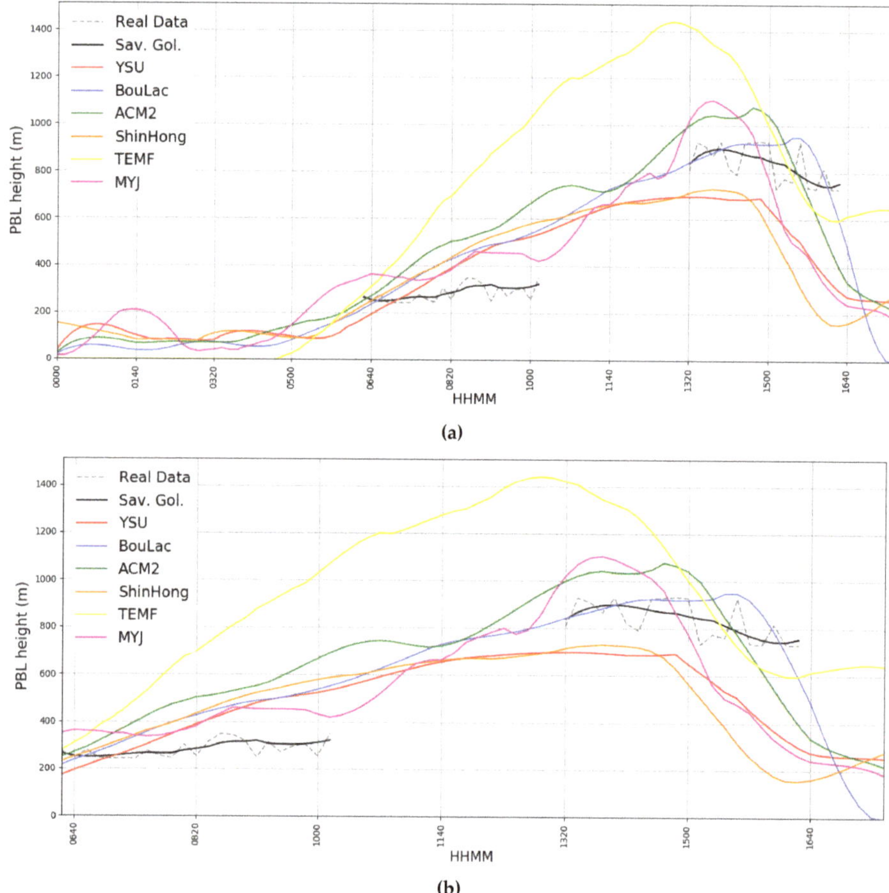

(a)

(b)

Figure 9. (**a**) Simulated PBLH timeseries compared with lidar-retrieved PBLH timeseries, 04/04/2016, dotted grey: real data, black: Savitsky–Golay smoothing of real data, red: simulated data with YSU parametrization, blue: simulated data with BouLac parametrization, green: simulated data with ACM2 parametrization, orange: simulated data with ShinHong parametrization, yellow: simulated data with TEMF parametrization, purple: simulated data with MYJ parametrization. (**b**) Simulated PBLH timeseries compared with lidar-retrieved PBLH timeseries, zoomed-in, 04/04/2016, dotted grey: real data, black: Savitsky–Golay smoothing of real data, red: simulated data with YSU parametrization, blue: simulated data with BouLac parametrization, green: simulated data with ACM2 parametrization, orange: simulated data with ShinHong parametrization, yellow: simulated data with TEMF parametrization, purple: simulated data with MYJ parametrization.

Simulated and observed temperature values are well correlated, and mean difference is under one degree Celsius in most cases (Figures 10 and 11) (Tables 1 and 2). The different simulations have similar behaviours with the exception of TEMF scheme that, in the 4 April episode, is unable to describe the temperature time trend.

The two events were characterised by low wind speed. During the first episode (Figure 12) the average wind speed was lower than 2 m/s at night and reached 3 m/s in the morning. All six PBL parametrizations correctly compute the low wind regime during the night and the increasing wind speed values from 4:00 UTC. The second episode (Figure 13) is characterised by a calm atmosphere with wind speeds less than 1.5 m/s. Also, in this case, all six simulations were able to identify the low wind regime. Therefore, MAE, MBE, and RMSE values for all simulations are generally lower than 1 m/s and a categorically-best parametrization cannot be found. Average wind speed is quite poorly correlated; however, such variation is to be partly expected, considering the low values of this average wind speed (Figures 12 and 13) (Tables 1 and 2).

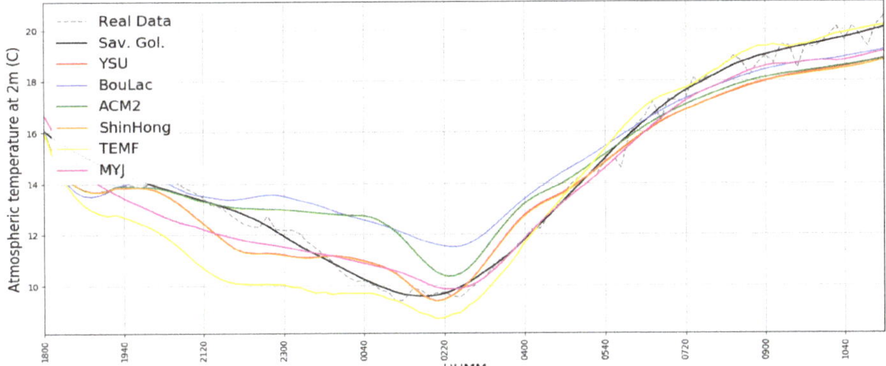

Figure 10. Simulated T2M timeseries compared with real GL T2M timeseries, 17/05/2017, dotted grey: real data, black: Savitsky–Golay smoothing of real data, red: simulated data with YSU parametrization, blue: simulated data with BouLac parametrization, green: simulated data with ACM2 parametrization, orange: simulated data with ShinHong parametrization, yellow: simulated data with TEMF parametrization, purple: simulated data with MYJ parametrization.

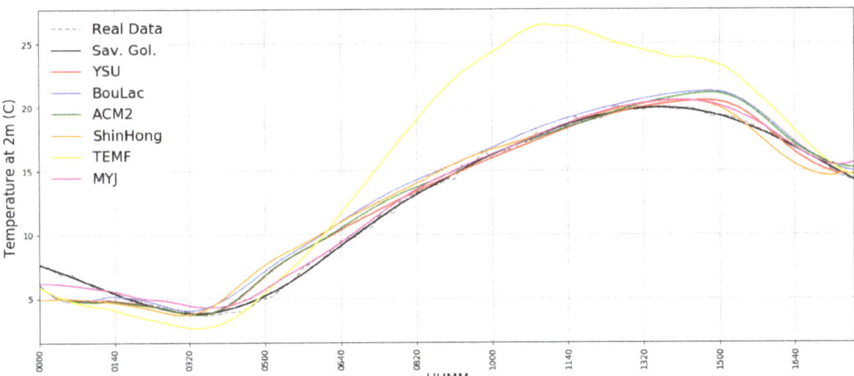

Figure 11. Simulated T2M timeseries compared with real GL T2M timeseries, 04/04/2016, dotted grey: real data, black: Savitsky–Golay smoothing of real data, red: simulated data with YSU parametrization, blue: simulated data with BouLac parametrization, green: simulated data with ACM2 parametrization, orange: simulated data with ShinHong parametrization, yellow: simulated data with TEMF parametrization, purple: simulated data with MYJ parametrization.

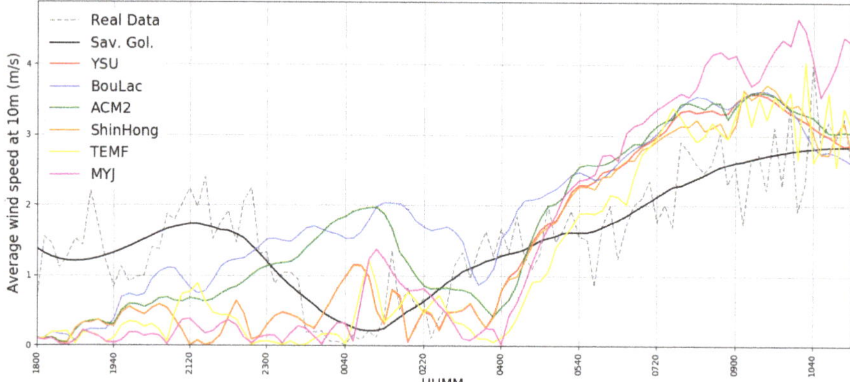

Figure 12. Simulated U10M timeseries compared with real GL U10M timeseries, 17/05/2017, dotted grey: real data, black: Savitsky–Golay smoothing of real data, red: simulated data with YSU parametrization, blue: simulated data with BouLac parametrization, green: simulated data with ACM2 parametrization, orange: simulated data with ShinHong parametrization, yellow: simulated data with TEMF parametrization, purple: simulated data with MYJ parametrization.

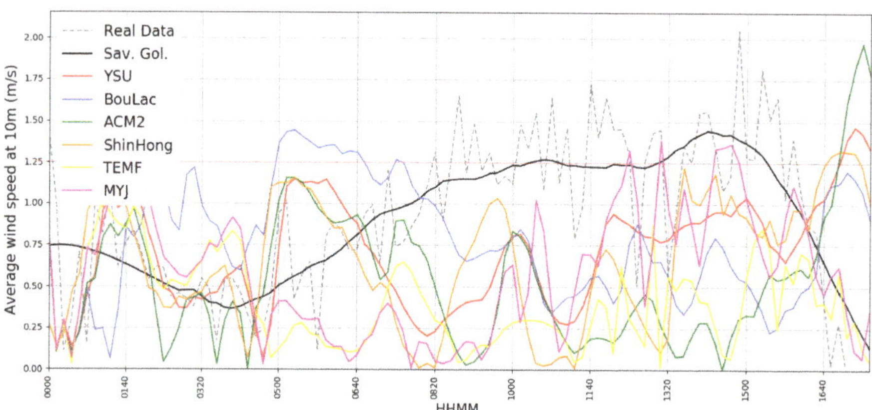

Figure 13. Simulated U10M timeseries compared with real GL U10M timeseries, 04/04/2016, dotted grey: real data, black: Savitsky–Golay smoothing of real data, red: simulated data with YSU parametrization, blue: simulated data with BouLac parametrization, green: simulated data with ACM2 parametrization, orange: simulated data with ShinHong parametrization, yellow: simulated data with TEMF parametrization, purple: simulated data with MYJ parametrization.

Pressure data is inconclusive; one case shows low correlation, another shows high correlation, yet both cases present relatively low mean difference compared to value evolution in time (Figures 14 and 15) (Tables 1 and 2). However, simulated pressure values from different parametrizations are very similar to one another. They differ from simulated ones regarding the trend, but the values are comparable. Statistical values of MAE, MBE and RMSE show a general common performance for all simulations.

The total precipitable water is consistently underestimated by all the WRF simulations by about 0.5 cm but the variation in time seems to be, however, appropriately predicted (Figures 16 and 17) (Tables 1 and 2). R^2 values show a decent correlation at best, because of a constantly significant mean difference and no interception between simulated and measured data. The above-mentioned Unified Noah Land Surface Model used in this simulation, judging from the original paper that introduces it [62], seems to show an increased sensitivity to variations of soil moisture, which might partly account for the difference in the total measured and simulated water vapor. However, such an error due to excessive sensitivity to soil moisture would also mean that the evolution of the total water vapor would be different from the simulated one: the fact that WRF manages to simulate quite accurately the evolution but not the quantity of total water vapor gives us reason to believe that the difference in simulated and observed values is due to unpredicted phenomena of water vapor transport and advection in the atmosphere. Another likely possibility is that these errors are caused by a dry bias in the initial or boundary conditions provided to WRF through the ECMWF data; a validation study has shown that instances of such a bias can appear for data over several complex terrains, which would certainly explain the differences between simulated and observed WV values [63].

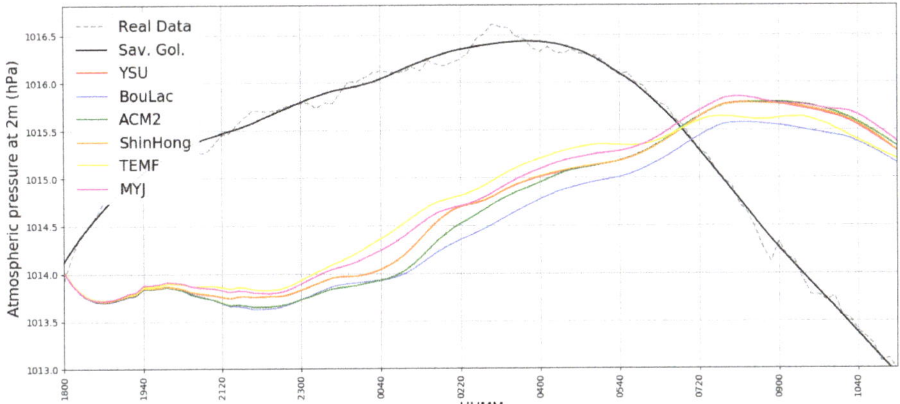

Figure 14. Simulated P2M timeseries compared with real GL P2M timeseries, 17/05/2017, dotted grey: real data, black: Savitsky–Golay smoothing of real data, red: simulated data with YSU parametrization, blue: simulated data with BouLac parametrization, green: simulated data with ACM2 parametrization, orange: simulated data with ShinHong parametrization, yellow: simulated data with TEMF parametrization, purple: simulated data with MYJ parametrization.

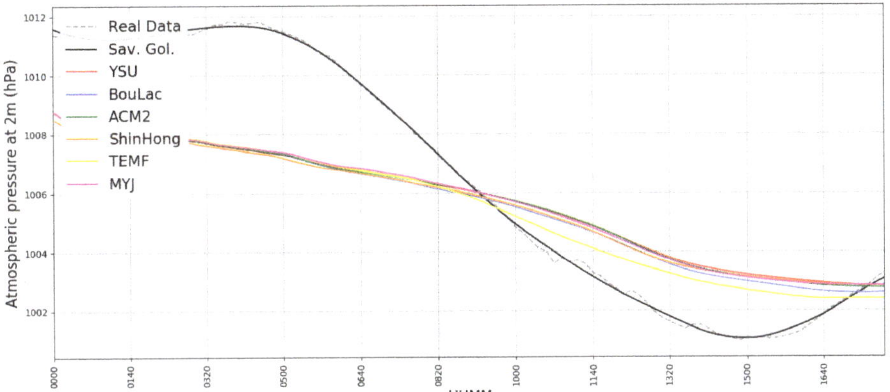

Figure 15. Simulated P2M timeseries compared with real GL P2M timeseries, 04/04/2016, dotted grey: real data, black: Savitsky–Golay smoothing of real data, red: simulated data with YSU parametrization, blue: simulated data with BouLac parametrization, green: simulated data with ACM2 parametrization, orange: simulated data with ShinHong parametrization, yellow: simulated data with TEMF parametrization, purple: simulated data with MYJ parametrization.

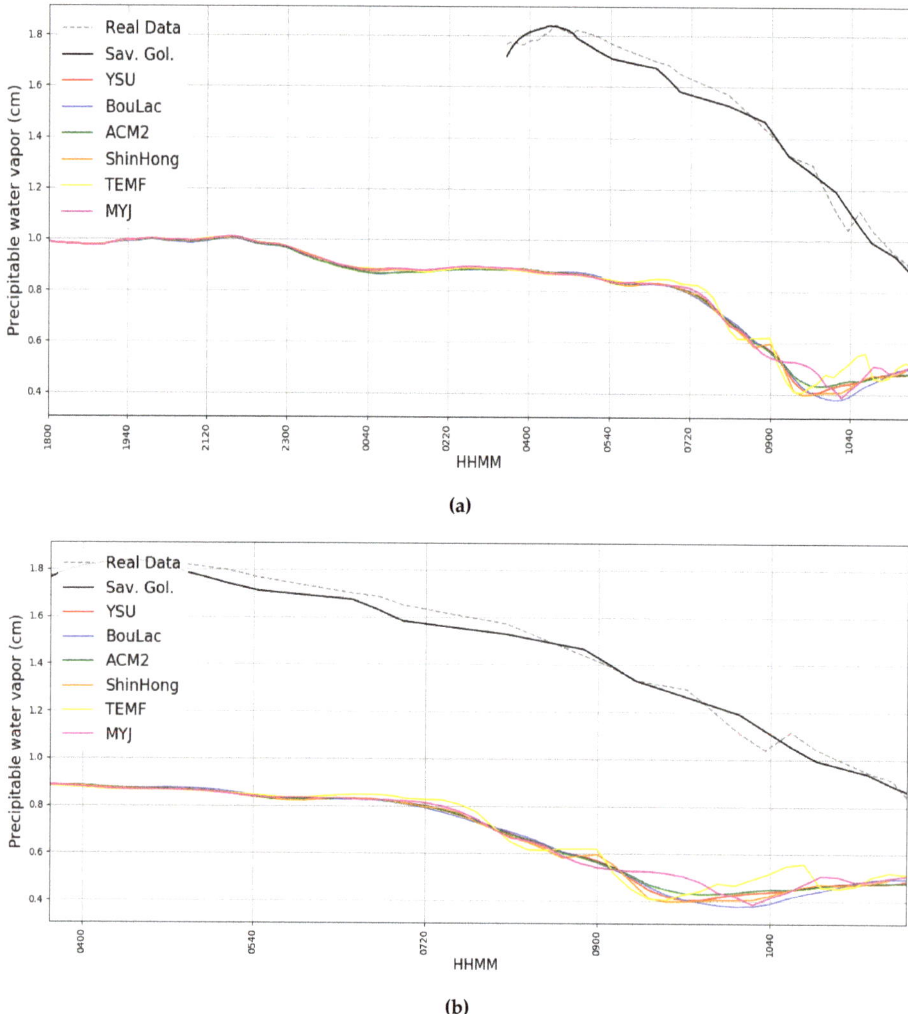

(a)

(b)

Figure 16. (**a**) Simulated WV timeseries compared with photometer-retrieved WV timeseries, 17/05/2017, dotted grey: real data, black: Savitsky–Golay smoothing of real data, red: simulated data with YSU parametrization, blue: simulated data with BouLac parametrization, green: simulated data with ACM2 parametrization, orange: simulated data with ShinHong parametrization, yellow: simulated data with TEMF parametrization, purple: simulated data with MYJ parametrization. (**b**) Simulated WV timeseries compared with photometer-retrieved WV timeseries, zoomed-in, 17/05/2017, dotted grey: real data, black: Savitsky–Golay smoothing of real data, red: simulated data with YSU parametrization, blue: simulated data with BouLac parametrization, green: simulated data with ACM2 parametrization, orange: simulated data with ShinHong parametrization, yellow: simulated data with TEMF parametrization, purple: simulated data with MYJ parametrization.

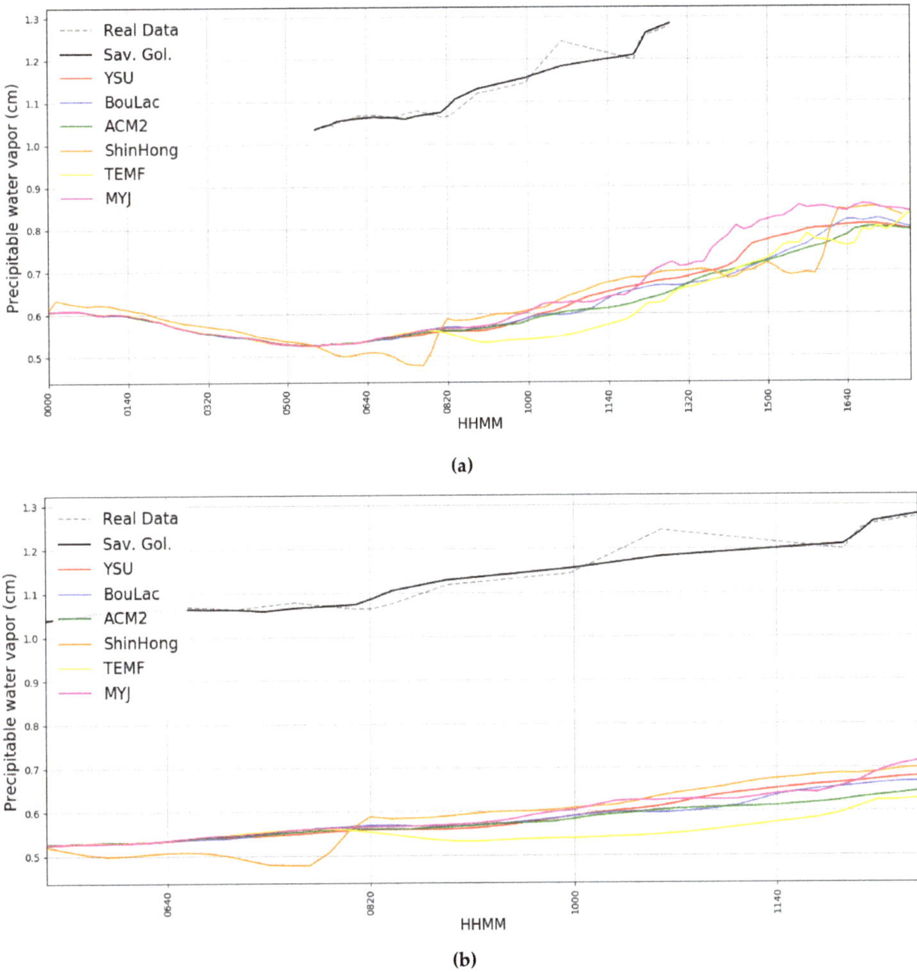

(a)

(b)

Figure 17. (**a**) Simulated WV timeseries compared with photometer-retrieved WV timeseries, 04/04/2016, dotted grey: real data, black: Savitsky–Golay smoothing of real data, red: simulated data with YSU parametrization, blue: simulated data with BouLac parametrization, green: simulated data with ACM2 parametrization, orange: simulated data with ShinHong parametrization, yellow: simulated data with TEMF parametrization, purple: simulated data with MYJ parametrization. (**b**) Simulated WV timeseries compared with photometer-retrieved WV timeseries, zoomed-in, 04/04/2016, dotted grey: real data, black: Savitsky–Golay smoothing of real data, red: simulated data with YSU parametrization, blue: simulated data with BouLac parametrization, green: simulated data with ACM2 parametrization, orange: simulated data with ShinHong parametrization, yellow: simulated data with TEMF parametrization, purple: simulated data with MYJ parametrization.

Table 1. Statistics of simulated PBLH, T2M, U10M, P2M and WV data compared to instrument-retrieved PBLH, T2M, U10M, P2M and WV data, 17/05/2017; performance indicators are MAE, MBE, RMSE, and R^2.

YSU		BouLac		ACM2		ShinHong		TEMF		MYJ	
PBLH		**PBLH**		**PBLH**		**PBLH**		**PBLH**		**PBLH**	
MAE	0.128	MAE	0.145	MAE	0.185	MAE	0.12	MAE	0.16	MAE	0.098
MBE	0.158	MBE	0.182	MBE	0.236	MBE	0.15	MBE	0.209	MBE	0.129
RMSE	0.302	RMSE	0.345	RMSE	0.444	RMSE	0.284	RMSE	0.397	RMSE	0.25
R^2	0.98	R^2	0.981	R^2	0.986	R^2	0.982	R^2	0.974	R^2	0.981
T2M		**T2M**		**T2M**		**T2M**		**T2M**		**T2M**	
MAE	0.713	MAE	1.011	MAE	0.898	MAE	0.707	MAE	0.89	MAE	0.556
MBE	0.048	MBE	0.082	MBE	0.069	MBE	0.048	MBE	0.068	MBE	0.038
RMSE	0.055	RMSE	0.11	RMSE	0.094	RMSE	0.055	RMSE	0.094	RMSE	0.046
R^2	0.971	R^2	0.948	R^2	0.954	R^2	0.971	R^2	0.949	R^2	0.98
U10M		**U10M**		**U10M**		**U10M**		**U10M**		**U10M**	
MAE	0.713	MAE	0.784	MAE	0.779	MAE	0.707	MAE	0.716	MAE	0.999
MBE	0.608	MBE	0.967	MBE	0.929	MBE	0.602	MBE	0.593	MBE	0.77
RMSE	0.859	RMSE	1.796	RMSE	1.826	RMSE	0.849	RMSE	0.824	RMSE	1.051
R^2	0.631	R^2	0.371	R^2	0.448	R^2	0.618	R^2	0.696	R^2	0.658
P2M		**P2M**		**P2M**		**P2M**		**P2M**		**P2M**	
MAE	1.472	MAE	1.521	MAE	1.526	MAE	1.45	MAE	1.347	MAE	1.437
MBE	0.015	MBE	0.015	MBE	0.015	MBE	0.014	MBE	0.013	MBE	0.014
RMSE	0.016	RMSE	0.016	RMSE	0.016	RMSE	0.015	RMSE	0.014	RMSE	0.015
R^2	0.105	R^2	0.145	R^2	0.135	R^2	0.103	R^2	0.036	R^2	0.094
WV		**WV**		**WV**		**WV**		**WV**		**WV**	
MAE	0.735	MAE	0.734	MAE	0.735	MAE	0.737	MAE	0.73	MAE	0.734
MBE	0.457	MBE	0.456	MBE	0.457	MBE	0.459	MBE	0.453	MBE	0.457
RMSE	0.464	RMSE	0.464	RMSE	0.464	RMSE	0.466	RMSE	0.461	RMSE	0.464
R^2	0.842	R^2	0.871	R^2	0.843	R^2	0.831	R^2	0.727	R^2	0.797

Strictly in terms of the MAE, for the 2017 data set, the MYJ parametrization obtained the best results with the PBLH and T2M evolution whereas much of the simulated data using the TEMF parametrization is highly inaccurate when compared to the PBLH and T2M sets of data obtained in 2016 (Figures 9 and 11, Table 2). In the case of P2M and WV, TEMF shows the smallest error, while the ShinHong parametrization is superior in the case of U10M (Tables 1 and 2). For the 2016 data set, T2M and U10M data are best modelled with MYJ, while TEMF and ShinHong are best in terms of the P2M and WV evolution respectively, and the BouLac parametrization obtains the best results regarding the PBLH evolution (Tables 1 and 2). YSU also yields satisfactory simulations in most cases, while the ACM2 parametrization seems to obtain consistently poor results.

Note that there are two instances when WRF simulations present consistent errors when compared to observed data: in both episodes, PBLH evolution is overestimated in the morning, and in both episodes, WV is underestimated (despite the fact that the evolution seems correctly predicted). There is also the problem with apparent simulation anomalies given when the model employs the TEMF parametrization. Numerous similarities can be identified between this parametrization and all the others, such as the order and non-locality; however, what this scheme seems to have as a substantial difference is the calculation of PBLH through a *Rib* method with zero as a threshold value. It is possible that this method is problematic for the simulations at hand and is unsuitable in the analyzed cases.

Table 2. Statistics of simulated PBLH, T2M, U10M, P2M and WV data compared to instrument-retrieved PBLH, T2M, U10M, P2M and WV data, 04/04/2016; performance indicators are MAE, MBE, RMSE, and R^2.

YSU		BouLac		ACM2		ShinHong		TEMF		MYJ	
PBLH		**PBLH**		**PBLH**		**PBLH**		**PBLH**		**PBLH**	
MAE	0.198	MAE	0.175	MAE	0.227	MAE	0.219	MAE	0.399	MAE	0.188
MBE	-0.057	MBE	0.094	MBE	0.226	MBE	-0.021	MBE	0.765	MBE	0.019
RMSE	0.411	RMSE	0.439	RMSE	0.549	RMSE	0.487	RMSE	1.088	RMSE	0.431
R^2	0.906	R^2	0.826	R^2	0.829	R^2	0.933	R^2	0.891	R^2	0.912
T2M		**T2M**		**T2M**		**T2M**		**T2M**		**T2M**	
MAE	0.616	MAE	1.044	MAE	0.761	MAE	0.909	MAE	3.522	MAE	0.411
MBE	0.074	MBE	0.112	MBE	0.084	MBE	0.113	MBE	0.27	MBE	0.048
RMSE	0.119	RMSE	0.152	RMSE	0.126	RMSE	0.17	RMSE	0.303	RMSE	0.067
R^2	0.98	R^2	0.976	R^2	0.978	R^2	0.963	R^2	0.933	R^2	0.995
U10M		**U10M**		**U10M**		**U10M**		**U10M**		**U10M**	
MAE	0.472	MAE	0.555	MAE	0.624	MAE	0.493	MAE	0.576	MAE	0.456
MBE	1.184	MBE	0.838	MBE	1.528	MBE	0.653	MBE	0.613	MBE	0.739
RMSE	4.777	RMSE	1.185	RMSE	6.374	RMSE	1.052	RMSE	0.672	RMSE	1.938
R^2	0.45	R^2	0.329	R^2	0.509	R^2	0.37	R^2	0.217	R^2	0.25
P2M		**P2M**		**P2M**		**P2M**		**P2M**		**P2M**	
MAE	2.249	MAE	2.160	MAE	2.233	MAE	2.241	MAE	1.998	MAE	2.199
MBE	0.022	MBE	0.021	MBE	0.022	MBE	0.022	MBE	0.02	MBE	0.022
RMSE	0.025	RMSE	0.025	RMSE	0.025	RMSE	0.026	RMSE	0.024	RMSE	0.025
R^2	0.922	R^2	0.93	R^2	0.918	R^2	0.939	R^2	0.952	R^2	0.929
WV		**WV**		**WV**		**WV**		**WV**		**WV**	
MAE	0.48	MAE	0.491	MAE	0.502	MAE	0.46	MAE	0.54	MAE	0.474
MBE	0.508	MBE	0.505	MBE	0.506	MBE	0.526	MAE	0.508	MBE	0.336
RMSE	0.508	RMSE	0.505	RMSE	0.506	RMSE	0.527	RMSE	0.509	RMSE	0.509
R^2	0.76	R^2	0.729	R^2	0.727	R^2	0.714	R^2	0.661	R^2	0.722

In order to test this possibility, two series of representative *Rib* profiles have been made using data simulated for the 2016 episode to test if the behaviour of the Rib profile made with the TEMF parametrization is also anomalous (Figures 18 and 19). These two series correspond to the two lidar measurement intervals in the 2016 episode, and the data has been picked so as to roughly match the middle of both of the lidar measurement intervals. The profiles are made as per the typical bulk Richardson number profile, using wind speed and potential temperature WRF output data [64]; they have not been smoothed with a Savitsky–Golay filter in order to show with a greater deal of precision the altitudes at which these profiles reach Ribc. As expected, the profiles do show that TEMF produces a Rib profile that is vastly different from the others; one would logically expect that a lower Ribc would produce a lower estimation of the PBLH, in opposition with the simulated PBLH values at hand. However, what is instead observed is that the TEMF-produced Rib profile remains negative at higher altitudes, approaching positive values at a much slower rate (Figures 18 and 19). Thus, the critical TEMF Rib = 0 is reached at higher altitudes than in the case of other parametrizations, resulting in an innacurate or anomalous simulated PBLH. Another interesting conclusion that can be extracted from the figures is that the Rib values simulated with all parametrizations become quite similar to one another at higher altitudes, except for TEMF (Figures 18 and 19).

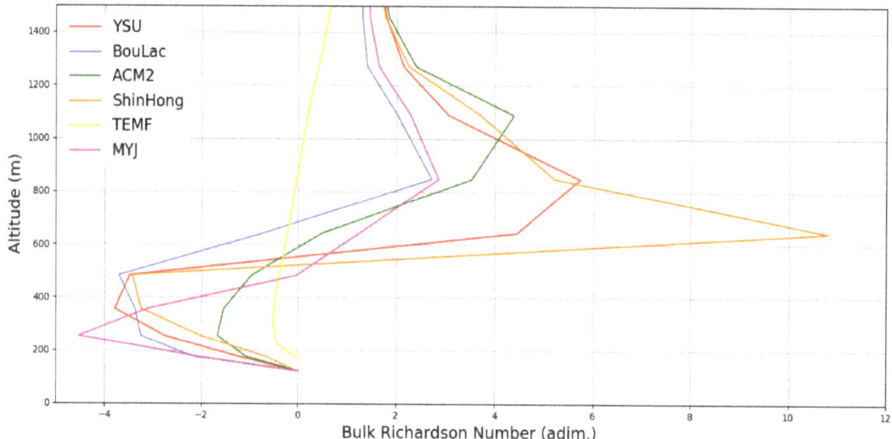

Figure 18. Simulated bulk Richardson number profiles compared with one another, 04/04/2016, 09:30, red: simulated data with YSU parametrization, blue: simulated data with BouLac parametrization, green: simulated data with ACM2 parametrization, orange: simulated data with ShinHong parametrization, yellow: simulated data with TEMF parametrization, purple: simulated data with MYJ parametrization.

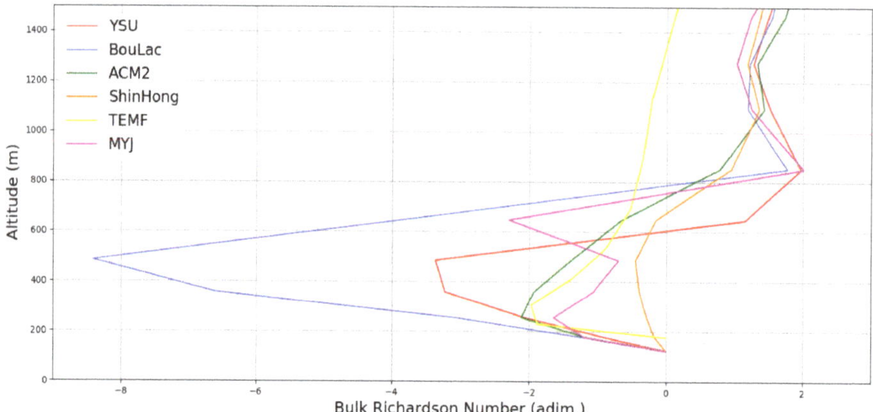

Figure 19. Simulated bulk Richardson number profiles compared with one another, 04/04/2016, 14:00, red: simulated data with YSU parametrization, blue: simulated data with BouLac parametrization, green: simulated data with ACM2 parametrization, orange: simulated data with ShinHong parametrization, yellow: simulated data with TEMF parametrization, purple: simulated data with MYJ parametrization.

6. Conclusion

Different PBL parametrization schemes in WRF framework were applied over a complex area crossed by rivers and including urban and rural zones. The simulated data were compared with observations from traditional surface instrumentation and a lidar system. A future study, facilitated by a suitable computational platform might include more sets of simulated data across multiple seasons, and a greater degree of experimentation with parametrizations pertaining to multiple aspects of the atmosphere.

The application of WRF in the selected timeframes and the selected region seems to yield, with most PBLH parametrizations, a boundary layer that is deeper in some cases from the obtained observed data. PBLH evolution is, on average, quite well modelled with the MYJ and also with YSU performing decently, with the exception of slight overestimations at the early hours of the episodes. Temperature seems to be the selected value that is consistently well-approximated. Wind speed regime is well represented in the case of low intensities, and this is done in both simulation episodes. Pressure values are equally well identified by all parametrizations, but the trend is not always correctly identified.

Given the various seasonal peculiarities, and the challenging geographic and anthropogenic complexity of the examined region, it can be said that the model performed well in the case of most chosen parameters; in the case of ground-level temperature, for example, studies show that differences of two to three degrees Celsius between WRF simulated and real data are not uncommon; nor are differences of more than 0.3 km in the case of PBL height [3–6]. These studies also seem to point out that such differences can increase in warm seasons. The comparisons show that multiple improvements can be made to WRF parametrizations for the specific region of Iași in the warm season, especially regarding the transport of precipitable water vapour in the atmosphere, and the early-morning evolution of the PBL. Additionally, simulations using the TEMF parametrization may present certain anomalous values, which most likely arise from the PBLH calculation that uses zero as the threshold in its Rib method.

Author Contributions: Conceptualization, R.L.-A., F.S. and C.M.M.; Data curation, R.L.-A., F.S., R.I. and C.M.M.; methodology, R.L.-A., F.S. and C.M.M.; software, R.L.-A. and F.S.; Supervision, R.L.-A., F.S. and C.M.M.; validation, R.L.-A. and F.S.; Visualization, F.S. and C.M.M.; formal analysis, R.L.-A., F.S., R.I., C.V. and C.M.M.; resources, R.L.-A., F.S., C.V. and C.M.M.; writing—original draft preparation, R.L.-A., F.S. and C.M.M.; writing—review and editing, R.L.-A., F.S., R.I., C.V. and C.M.M.; project administration, R.L.-A., F.S. and C.M.M.; funding acquisition, C.M.M.; Investigation, R.L.-A., F.S., R.I., C.V. and C.M.M.

Funding: This work was supported by a research grant of the TUIASI, project number GnaC2018_65/2019.

Conflicts of Interest: The authors declare no conflict of interest.

References

1. Rao, K.S. Uncertainty analysis in atmospheric dispersion modeling. *Pure Appl. Geophys.* **2005**, *162*, 1893–1917. [CrossRef]

2. Selvam, A.M. Nonlinear dynamics and chaos: Applications in atmospheric sciences. *J. Adv. Math. Appl.* **2012**, *1*, 181–205. [CrossRef]

3. Timofte, A.; Belegante, L.; Cazacu, M.M.; Albina, B.; Talianu, C.; Gurlui, S. Study of planetary boundary layer height from LIDAR measurements and ALARO model. *J. Optoelectron. Adv. Mater.* **2015**, *17*, 911–917.

4. Banks, R.F.; Tiana-Alsina, J.; Baldasano, J.M.; Rocadenbosch, F.; Papayannis, A.; Solomos, S.; Tzanis, C.G. Sensitivity of boundary-layer variables to PBL schemes in the WRF model based on surface meteorological observations, lidar, and radiosondes during the HygrA-CD campaign. *Atmos. Res.* **2016**, *176*, 185–201. [CrossRef]

5. Belegante, L.; Nicolae, D.; Nemuc, A.; Talianu, C.; Derognat, C. Retrieval of the boundary layer height from active and passive remote sensors. Comparison with a NWP model. *Acta Geophys.* **2014**, *62*, 276–289. [CrossRef]

6. García-Díez, M.; Fernández, J.; Fita, L.; Yagüe, C. Seasonal dependence of WRF model biases and sensitivity to PBL schemes over Europe. *Q. J. R. Meteorol. Soc.* **2013**, *139*, 501–514. [CrossRef]

7. Banks, R.F.; Tiana-Alsina, J.; Rocadenbosch, F.; Baldasano, J.M. Performance evaluation of the boundary-layer height from lidar and the Weather Research and Forecasting model at an urban coastal site in the north-east Iberian Peninsula. *Bound. Layer Meteorol.* **2015**, *157*, 265–292. [CrossRef]

8. Boadh, R.; Satyanarayana, A.N.V.R.; Krishna, T.V.B.P.S.; Madala, S. Sensitivity of PBL schemes of the WRF-ARW model in simulating the boundary layer flow parameters for their application to air pollution dispersion modelling over a tropical station. *Atmosfera* **2016**, *29*, 61–81. [CrossRef]

9. Milovac, J.; Warrach-Sagi, K.; Behrendt, A.; Späth, F.; Ingwersen, J.; Wulfmeyer, V. Inversigation of PBL schemes combining the WRF model simulations with scanning water vapor differential abseorption lidar measurements. *J. Geophys. Res. Atmos.* **2015**, *121*, 624–649. [CrossRef]

10. Sathyanadh, A.; Prabha, T.V.; Balaji, B.; Resmi, E.A.; Karipot, A. Evaluation of WRF PBL parametrization schemes against direct observations during a dry event over the Ganges valley. *Atmos. Res.* **2017**, *193*, 125–141. [CrossRef]

11. Pérez, C.; Jiménez, P.; Jorba, O.; Sicard, M.; Baldasano, J.M. Influence of the PBL scheme on high-resolution photochemical simulations in an urban coastal area over the Western Mediterranean. *Atmos. Environ.* **2006**, *40*, 5274–5297. [CrossRef]

12. Bossioli, E.; Tombrou, M.; Dandou, A.; Athanasopoulou, E.; Varotsos, K.V. The role of planetary boundary-layer parameterizations in the air quality of an urban area with complex topography. *Bound. Layer Meteorol.* **2009**, *131*, 53–72. [CrossRef]

13. Dudhia, J. A nonhydrostatic version of the Penn State–NCAR mesoscale model: Validation tests and simulation of an Atlantic cyclone and cold front. *Mon. Weather Rev.* **1993**, *121*, 1493–1513. [CrossRef]

14. Lupașcu, A.; Iriza, A.; Dumitrache, R.C. Using a high resolution topographic data set and analysis of the impact on the forecast of meteorological parameters. *Rom. Rep. Phys.* **2015**, *67*, 653–664.

15. Iriza, A.; Dumitrache, R.C.; Ștefan, S. Numerical modelling of the Bucharest urban heat island with the WRF-urban system. *Rom. J. Phys.* **2017**, *62*, 1–14.

16. Isvoranu, D.; Badescu, V. Comparison Between Measurements and WRF Numerical Simulation of Global Solar Irradiation in Romania. *Ann. West. Univ. Timis. Phys.* **2013**, *57*, 24–33. [CrossRef]

17. Dimitrova, R.; Danchovski, V.; Egova, E.; Vladimirov, E.; Sharma, A.; Gueorguiev, O.; Ivanov, D. Modeling the Impact of Urbanization on Local Meteorological Conditions in Sofia. *Atmosphere* **2019**, *10*, 366. [CrossRef]

18. Run, R.; Case, R.D.T. User's Guide for the NMM Core of the Weather Research and Forecast (WRF) Modeling System Version 3. Available online: https://dtcenter.org/wrf-nmm/users/docs/user_guide/V3/contents_nmm.pdf (accessed on 15 June 2019).

19. Morille, Y.; Haeffelin, M.; Drobinski, P.; Pelon, J. STRAT: An automated algorithm to retrieve the vertical structure of the atmosphere from single-channel lidar data. *J. Atmos. Ocean. Technol.* **2007**, *24*, 761–775. [CrossRef]

20. Hennemuth, B.; Lammert, A. Determination of the atmospheric boundary layer height from radiosonde and lidar backscatter. *Bound. Layer Meteorol.* **2006**, *120*, 181–200. [CrossRef]

21. Weather Research and Forecasting Model. Available online: https://www.mmm.ucar.edu/weather-research-and-forecasting-model (accessed on 10 May 2019).

22. Powers, J.G.; Klemp, J.B.; Skamarock, W.C.; Davis, C.A.; Dudhia, J.; Gill, D.O.; Grell, G.A. The weather research and forecasting model: Overview, system efforts, and future directions. *Bull. Am. Meteorol. Soc.* **2017**, *98*, 1717–1737. [CrossRef]

23. Skamarock, W.C.; Klemp, J.B.; Dudhia, J.; Gill, D.O.; Barker, D.M.; Wang, W.; Powers, J.G. *A Description of the Advanced Research WRF Version 2*; No. NCAR/TN-468+ STR; National Center for Atmospheric Research: Boulder, Co, USA; Mesoscale and Microscale Meteorology Div.: Boulder, CO, USA, 2015.

24. Jiménez, P.A.; Dudhia, J.; González-Rouco, J.F.; Montávez, J.P.; García-Bustamante, E.; Navarro, J.; Muñoz-Roldán, A. An evaluation of WRF's ability to reproduce the surface wind over complex terrain based on typical circulation patterns. *J. Geophys. Res. Atmos.* **2013**, *118*, 7651–7669. [CrossRef]

25. Geiger, R. *Klassifikation Der Klimate Nach W. Köppen (Classification of Climates after W. Köppen)*; Landolt-Börnstein—Zahlenwerte und Funktionen aus Physik, Chemie, Astronomie, Geophysik und Technik, alte Serie; Springer: Berlin, Germany, 1954; Volume 3, pp. 603–607.

26. Peel, M.C.; Finlayson, B.L.; McMahon, T.A. Updated world map of the Köppen–Geiger climate classification. *Hydrol. Earth Syst. Sci.* **2007**, *11*, 1633–1644. [CrossRef]

27. Pisciculture Tiganasi. Available online: https://www.producator-agricol.ro/piscicultura/tiganasi (accessed on 1 July 2019).

28. Stull, R.B. *An Introduction to Boundary Layer Meteorology*; Kluwer Academic Publishers: Dordrecht, The Netherlands; Boston, MA, USA; London, UK, 1988; p. 666.

29. Foken, T. *Micrometeorolgy*; Springer-Verlag: Berlin, Germany, 2016. [CrossRef]

30. World Population Review. Available online: http://worldpopulationreview.com/countries/romania-population/cities/ (accessed on 1 July 2019).

31. Aspen Economic Opportunities & Financing the Economy Program 2018. Available online: http://aspeninstitute.ro/white-paper-aspen-economic-opportunities-financing-the-economy-program-2018/ (accessed on 1 July 2019).

32. AirVisual Database. Available online: https://www.airvisual.com/world-most-polluted-cities (accessed on 15 May 2019).

33. Hong, S.Y.; Lim, J.O.J. The WRF single-moment 6-class microphysics scheme. *Asia-Pac. J. Atmos. Sci.* **2006**, *42*, 129–151.

34. Janić, Z.I. Nonsingular implementation of the Mellor-Yamada level 2.5 scheme in the NCEP Meso model. *Natl. Ocean. Atmos. Adm.* **2001**, *437*, 1–61.

35. Pleim, J.E. A combined local and nonlocal closure model for the atmospheric boundary layer. Part I: Model description and testing. *J. Appl. Meteorol. Climatol.* **2007**, *46*, 1383–1395. [CrossRef]

36. Bougeault, P.; Lacarrere, P. Parameterization of orography-induced turbulence in a mesobeta–scale model. *Mon. Weather Rev.* **1989**, *117*, 1872–1890. [CrossRef]

37. Angevine, W.M.; Jiang, H.; Mauritsen, T. Performance of an eddy diffusivity–mass flux scheme for shallow cumulus boundary layers. *Mon. Weather Rev.* **2010**, *138*, 2895–2912. [CrossRef]

38. Shin, H.H.; Hong, S.Y. Analysis of resolved and parameterized vertical transports in convective boundary layers at gray-zone resolutions. *J. Atmos. Sci.* **2013**, *70*, 3248–3261. [CrossRef]

39. Martilli, A.; Clappier, A.; Rotach, M.W. An urban surface exchange parameterisation for mesoscale models. *Bound.-Layer Meteorol.* **2002**, *104*, 261–304. [CrossRef]

40. Durre, I.; Vose, R.S.; Wuertz, D.B. Robust automated quality assurance of radiosonde temperatures. *J. Appl. Meteorol. Climatol.* **2008**, *47*, 2081–2095. [CrossRef]

41. Kropfli, R.A. A review of microwave radar observations in the dry convective planetary boundary layer. *Bound.-Layer Meteorol.* **1983**, *26*, 51–67. [CrossRef]

42. Beyrich, F.; Görsdorf, U. Composing the diurnal cycle of mixing height from simultaneous sodar and wind profiler measurements. *Bound.-Layer Meteorol.* **1995**, *76*, 387–394. [CrossRef]

43. Cimini, D.; De Angelis, F.; Dupont, J.C.; Pal, S.; Haeffelin, M. Mixing layer height retrievals by multichannel microwave radiometer observations. *Atmos. Meas.* **2013**, *6*, 2941–2951. [CrossRef]

44. Rosu, I.A.; Cazacu, M.M.; Prelipceanu, O.S.; Agop, M. A Turbulence-Oriented Approach to Retrieve Various Atmospheric Parameters Using Advanced Lidar Data Processing Techniques. *Atmosphere* **2019**, *10*, 38. [CrossRef]

45. Lolli, S.; Madonna, F.; Rosoldi, M.; Campbell, J.R.; Welton, E.J.; Lewis, J.R.; Pappalardo, G.; Gu, Y. Impact of varying lidar measurement and data processing techniques in evaluating cirrus cloud and aerosol direct radiative effects. *Atmos. Meas. Tech.* **2018**, *11*, 1639–1651. [CrossRef]

46. Reichardt, J.; Wandinger, U.; Klein, V.; Mattis, I.; Hilber, B.; Begbie, R. RAMSES: German Meteorological Service autonomous Raman lidar for water vapor, temperature, aerosol, and cloud measurements. *Appl. Opt.* **2012**, *51*, 8111–8131. [CrossRef] [PubMed]

47. Belegante, L.; Cazacu, M.M.; Timofte, A.; Toanca, F.; Vasilescu, J.; Rusu, M.I.; Ajtai, N.; Stefanie, H.I.; Vetres, I.; Ozunu, A.; et al. Case study of the first volcanic ash exercise in Romania using remote sensing techniques. *Environ. Eng. Manag. J.* **2015**, *14*, 2503–2504.

48. Papayannis, A.; Nicolae, D.; Kokkalis, P.; Binietoglou, I.; Talianu, C.; Belegante, L.; Tsaknakis, G.; Cazacu, M.M.; Vetres, I.; Ilic, L. Optical, size and mass properties of mixed type aerosols in Greece and Romania as observed by synergy of lidar and sunphotometers in combination with model simulations: A case study. *Sci. Total Environ.* **2014**, *500*, 277–294. [CrossRef] [PubMed]

49. Flamant, C.; Pelon, J.; Flamant, P.H.; Durand, P. Lidar determination of the entrainment zone thickness at the top of the unstable marine atmospheric boundary layer. Bound. *Layer Meteorol.* **1997**, *83*, 247–284. [CrossRef]

50. Haeffelin, M.; Angelini, F.; Morille, Y.; Martucci, G.; Frey, S.; Gobbi, G.P.; Lolli, S.; O'Dowd, C.D.; Sauvage, L.; Xueref-Rémy, I.; et al. Evaluation of mixing-height retrievals from automatic profiling lidars and ceilometers in view of future integrated networks in Europe. Bound. *Layer Meteorol.* **2012**, *143*, 49–75. [CrossRef]

51. Klett, J.D. Stable analytical inversion solution for processing lidar returns. *Appl. Opt.* **1981**, *20*, 211–220. [CrossRef] [PubMed]

52. Liu, Z.; Hunt, W.; Vaughan, M.; Hostetler, C.; McGill, M.; Powell, K.; Winker, D.M.; Hu, Y. Estimating random errors due to shot noise in backscatter lidar observations. *Appl. Opt.* **2006**, *45*, 4437–4447. [CrossRef] [PubMed]

53. *Hamamatsu, Photomultiplier Tubes, and Photomultipliers Tubes Photonics "Basics and Applications"*; Hamamatsu Photonics KK: Iwata City, Japan, 2007.

54. Unga, F.; Cazacu, M.M.; Timofte, A.; Bostan, D.; Mortier, A.; Dimitriu, D.G.; Gurlui, S.; Goloub, P. Study of tropospheric aerosol types over Iasi, Romania, during summer of 2012. Environ. *Eng. Manag. J.* **2013**, *12*, 297–303.

55. Cazacu, M.M.; Timofte, A.; Unga, F.; Albina, B.; Gurlui, S. AERONET data investigation of the aerosol mixtures over Iasi area, One-year time scale overview. *J. Quant. Spectrosc. Radiat. Transf.* **2015**, 15357–15364. [CrossRef]

56. Ajtai, N.; Stefanie, H.; Arghius, V.; Meltzer, M.; Costin, D. Characterization of aerosol optical and microphysical properties over North-Western Romania in correlation with predominant atmospheric circulation patterns, International Multidisciplinary Scientific Geo Conference. *Surv. Geol. Min. Ecol. Manag.* **2017**, *17*, 375–382.

57. Ajtai, N.; Stefanie, H.; Ozunu, A. Description of aerosol properties over Cluj-Napoca derived from AERONET sun photometric data. *Environ. Eng. Manag. J.* **2013**, *12*, 227–232. [CrossRef]

58. Binietoglou, I.; Basart, S.; Alados-Arboledas, L.; Amiridis, V.; Argyrouli, A.; Baars, H.; Baldasano, J.M.; Balis, D.; Belegante, L.; Bravo-Aranda, J.A.; et al. A methodology for investigating dust model performance using synergistic EARLINET/AERONET dust concentration retrievals. *Atmos. Meas. Tech.* **2015**, *8*, 3577–3600. [CrossRef]

59. Steinier, J.; Termonia, Y.; Deltour, J. Smoothing and differentiation of data by simplified least square procedure. *Anal. Chem.* **1972**, *44*, 1906–1909. [CrossRef]

60. Savitzky, A.; Golay, M.J. Smoothing and differentiation of data by simplified least squares procedures. *Anal. Chem.* **1964**, *36*, 1627–1639. [CrossRef]

61. Glossary of Meteorology. Available online: http://glossary.ametsoc.org/wiki/Precipitable_water (accessed on 1 June 2019).

62. Tewari, M.; Chen, F.; Wang, W.; Dudhia, J.; LeMone, M.A.; Mitchell, K.; Cuenca, R.H. Implementation and verification of the unified NOAH land surface model in the WRF model. In *20th Conference on Weather Analysis and Forecasting/16th Conference on Numerical Weather Prediction*; American Meteorological Society: Seattle, WA, USA, 2004; Volume 1115.

63. Bock, O.; Keil, C.; Richard, E.; Flamant, C.; Bouin, M.N. Validation of precipitable water from ECMWF model analyses with GPS and radiosonde data during the MAP SOP. *Q. J. R. Meteorol. Soc. A J. Atmos. Sci. Appl. Meteorol. Phys. Oceanogr.* **2005**, *131*, 3013–3036. [CrossRef]

64. Zhang, Y.; Gao, Z.; Li, D.; Li, Y.; Zhang, N.; Zhao, X.; Chen, J. On the computation of planetary boundary-layer height using the bulk Richardson number method. *Geosci. Model. Dev.* **2014**, *7*, 2599–2611. [CrossRef]

Article

Multiyear Typology of Long-Range Transported Aerosols over Europe

Victor Nicolae [1,2], Camelia Talianu [1,3], Simona Andrei [1], Bogdan Antonescu [1,2], Dragoş Ene [1], Doina Nicolae [1], Alexandru Dandocsi [1,4], Victorin-Emilian Toader [5], Sabina Ştefan [2], Tom Savu [6] and Jeni Vasilescu [1,*]

[1] National Institute of Research and Development for Optoelectronics INOE2000, Atomistilor 409, RO77125 Magurele, Romania
[2] Faculty of Physics, University of Bucharest, Atomistilor 405, RO77125 Magurele, Romania
[3] Institute of Meteorology and Climatology, University of Natural Resources and Life Sciences, 1180 Vienna, Austria
[4] Faculty of Applied Sciences, Politehnica University of Bucharest, Splaiul Independenţei 313, RO060042 Bucureşti, Romania
[5] National Institute for Earth Physics, Calugareni 12, RO077125 Magurele, Romania
[6] DOLSAT Consult, Al. Valea lui Mihai 2, sector 6, RO061756 Bucureşti, Romania
* Correspondence: jeni@inoe.ro

Received: 16 July 2019; Accepted: 16 August 2019; Published: 22 August 2019

Abstract: In this study, AERONET (Aerosol Robotic Network) and EARLINET (European Aerosol Research Lidar Network) data from 17 collocated lidar and sun photometer stations were used to characterize the optical properties of aerosol and their types for the 2008–2018 period in various regions of Europe. The analysis was done on six cluster domains defined using circulation types around each station and their common circulation features. As concluded from the lidar photometer measurements, the typical aerosol particles observed during 2008–2018 over Europe were medium-sized, medium absorbing particles with low spectral dependence. The highest mean values for the lidar ratio at 532 nm were recorded over Northeastern Europe and were associated with Smoke particles, while the lowest mean values for the Angstrom exponent were identified over the Southwest cluster and were associated with Dust and Marine particles. Smoke (37%) and Continental (25%) aerosol types were the predominant aerosol types in Europe, followed by Continental Polluted (17%), Dust (10%), and Marine/Cloud (10%) types. The seasonal variability was insignificant at the continental scale, showing a small increase in the percentage of Smoke during spring and a small increase of Dust during autumn. The aerosol optical depth (AOD) slightly decreased with time, while the Angstrom exponent oscillated between "hot and smoky" years (2011–2015) on the one hand and "dusty" years (2008–2010) and "wet" years (2017–2018) on the other hand. The high variability from year to year showed that aerosol transport in the troposphere became more and more important in the overall balance of the columnar aerosol load.

Keywords: aerosol typing; lidar; photometry; EARLINET; AERONET

---

1. Introduction

Aerosols influence the planetary radiation budget in both direct and indirect ways by absorbing and scattering solar radiation or by changing cloud properties [1]. Knowledge of aerosol physical and chemical properties can increase our current level of understanding of aerosol effects, variability, and their emerging role in climate change. Global and regional datasets of aerosol properties can be obtained from satellite or ground-based measurements (e.g., References [2,3]). Satellite data give good spatial coverage but suffer from poor temporal coverage; however, this is currently compensated for by synergistic approaches using ground-based measurements. Continental ground-based networks such as the European Aerosol Research Lidar Network (EARLINET, https://www.earlinet.org) and global networks such as the Aerosol Robotic Network (AERONET, https://aeronet.gsfc.nasa.gov/) have measured aerosol properties for more than two decades, building on data quality through their complex quality assurance programs. These long-term databases give the opportunity to assess the main properties of aerosol types, their variability at least on a continental scale, the processes driving property changes, long-range transport, or seasonal occurrence (e.g., References [3,4]). Recently, the two networks have joined their efforts in the frame of the Aerosols, Clouds, and Trace gases Research InfraStructure (ACTRIS, https://www.actris.eu/, [5,6]).

Raman lidar systems are the basic instruments used on regular bases in EARLINET. The continental scale coverage and simultaneous measurements schedule give the opportunity to study aerosol properties using vertically resolved data [5,7,8]. Lidar systems can provide information regarding the aerosol content on multiple layers due to their temporal and vertical high resolution. A distinct classification of the predominant aerosol on each layer is thus possible when using the intensive lidar parameters, which can be retrieved when following the usual practice within the EARLINET community. The classification includes layer boundaries identification and the calculation of intensive optical property values (i.e., lidar ratios, Angstrom exponents, particle linear depolarization ratios) specific to the layer and evaluation, taking into account other instruments or model outputs [9,10]. This subjective approach is highly time-consuming, and a subjective mode to derive the aerosol types is usually conducted only for special cases.

AERONET is a global network formed by ground-based sun–sky or lunar photometers and was established more than 25 years ago by NASA (National Aeronautics and Space Administration, i.e., globally) and PHOTONS (PHOtométrie pour le Traitement Opérationnel de Normalisation Satellitaire, i.e., for Europe) [11,12]. Long-term optical, microphysical, and radiative columnar aerosol properties are retrieved in a standardized and controlled mode. Evaluation and classification of aerosols are possible using the key parameters retrieved from photometer data. Aerosol optical properties from AERONET data are also used on different classification methods to determine the predominant aerosol type [13,14]. Biomass burning or desert dust aerosols are some of the classes that can be identified due to their distinct optical parameters [15,16]. The extinction Angstrom exponent (AE) 380/500 nm values near unit are assumed to be characteristic for industrial aerosols, while there are smaller values for dust types and larger values for biomass burning aerosols [16]. Similarly, the high values of aerosol optical depth (AOD) at 500 nm correspond to biomass burning events, while near zero represents the background level [17].

The aerosol mixtures are difficult to assess from sun photometers or lidar retrievals because different aerosol types have overlapping characteristic parameters. The combination of collocated lidar and photometer data give the opportunity to better assess aerosol types. The parameters used to identify the aerosol types are often a combination of lidar ratios (LRs), particle depolarization ratios (PDRs), AEs, and AODs [8,18,19].

The accuracy of aerosol classification is low using a subjective approach due to both multiple mandatory assumptions and similar properties of different aerosol types. Most of the typing methods are focused on advanced lidar systems, such as high spectral resolution lidar [7] or multiwavelength Raman measurements [8], sometimes also including particle linear depolarization. More automatic approaches have been explored in recent studies dedicated to aerosol typing using optical parameters retrieved from ground-based or satellite lidar measurements [5,7,9,19,20].

In this study, both AERONET and EARLINET data were used to characterize the optical properties of aerosol for the 2008–2018 period in various regions of Europe. For EARLINET data, an artificial neural network (ANN) software developed by Nicolae D. et al. [20] was used to derive the predominant type of aerosol. The ANN tends to simulate the human neural network comportment, learning after multiple similar situations and being able to choose between comparable inputs. ANNs are widely used in data processing or classification, including applications in atmospheric sciences, due to their capacity to deal with noisy data and superposed clusters. The ANN capacity to deal with nonlinear relationships between data is exploited in lidar data processing [21], improvements in satellite data related to aerosol properties [22], and retrieval of atmospheric temperature profiles from satellite measurements [23].

The aim of this article is to provide for the first time, to the authors' knowledge, a multiyear typology of long-range transported aerosols over Europe. The analysis was made on the basis of lidar data from EARLINET and collocated sun photometer data from AERONET. The article is structured as follows: Section 2 describes the observational data used to do the aerosol typing over Europe, a neural network aerosol typing algorithm based on lidar data, and the defined regions based on typical air mass transport. Optical characteristics of aerosols and aerosol types over Europe are discussed in Section 3. Section 4 summarizes the concluding remarks.

2. Data and Methodology

To analyze the optical characteristics of the aerosols and their type, a combined set of collocated lidar and sun photometer data was used. Optical properties of aerosols were retrieved from lidar and separately from sun photometers and then compared. A neural network aerosol typing algorithm based on lidar data (NATALI) software, already tested on synthetic data and a small set of measured data [20], was used to classify aerosols using lidar data from EARLINET stations.

2.1. Observational Data

The EARLINET database (i.e., multispectral Raman lidar climatological data available from [24]) and the AERONET database (i.e., sun/sky photometer measurements available from [25]) were the two main data sources used in this article to obtain a multiyear typology of long-range transported aerosols over Europe between 2008 and 2018. Geographically distributed stations that operated both multiwavelength Raman lidars and automatic sun/sky photometers were selected, considering the availability of long-term records from both instruments (Figure 1 and Table 1).

The EARLINET climatological measurements were systematically performed every Monday at (locally) noon and every Monday and Thursday evening after local sunset. The measurements were performed for at least 2 h at the wavelengths 355 nm, 532 nm, and 1064 nm for the elastic channels, 387 nm and 607 nm for the Raman channels, and 532 nm for the depolarization channel. In order to assess the quality of the measurements, EARLINET developed a specific quality assurance program, which is mandatory for all lidar stations [5]. This program is improved year by year and is currently being developed by the Centre for Aerosol Remote Sensing in ACTRIS.

Combined method [26] and inversion algorithms [27,28] were applied to lidar measurements to retrieve the vertical profiles of the aerosol backscatter coefficient at 355 nm, 532 nm, and 1064 nm, and the aerosol extinction coefficient at 355 nm and 532 nm. The calibrated depolarization vertical profiles of the aerosols were computed using the method proposed by Freudenthaler [29] and Belegante et al. [30]. The inversion of lidar measurements to retrieve the aerosol backscatter and extinction coefficient profiles and their associated uncertainties was done for many years at the level of each lidar station using its own software (therefore an important vulnerability for the network). Only recently, EARLINET developed the single calculus chain (SCC) [31], a centralized software to process lidar data from all stations in a homogeneous and quality assured manner. However, the data presented in this paper were not processed with the SCC, which was not available at the time the measurements were done. Due to the small number of calibrated depolarization datasets, they were also not used.

Starting from the aerosol extensive properties (backscatter and extinction coefficient profiles), all subsequent analysis was made using the NATALI software [20]. Lidar datasets collected between 2008 and 2018 by 17 stations of EARLINET distributed over all of Europe (Figure 1) were used as input. The aerosol layers were identified from the lidar measurements with the gradient method based on the identification of the peaks and valleys from the first derivative applied to the vertical profiles of backscatter coefficients at 1064 nm. Multiple criteria regarding the quality assurance (QA) of lidar data were used to make the aerosol typing as accurate as possible. The NATALI software performs aerosol typing only if all aerosol intensive optical parameters are provided and their values are between the thresholds that are defined based on previous studies [20]. The typing process is done for layers that pass these criteria, and otherwise the probability of misclassifying or returning an unknown type increases. In addition, a flag stating that typing is uncertain appears when the relative error of any of the intensive optical parameters is higher than 20%. Besides these QA criteria already implemented in the NATALI software, an extra filter was applied to the EARLINET climatological datasets. Only intensive optical parameter sets that corresponded to possible physical characteristics of aerosols were retained for aerosol typing. Thus, aerosol layers characterized by small particles with low absorbance or by big particles and high absorbance were considered physically highly improbable and were removed from the typing datasets. After all QA criteria were applied to the total number of 4957 layers from lidar, only 808 were selected for aerosol typing (~16%) (Table 1).

The AERONET photometer measurements of the solar and sky radiance were performed automatically at 340 nm, 380 nm, 440 nm, 500 nm, 670 nm, 870 nm, and 1020 nm. The columnar optical properties (i.e., aerosol optical depth, Angstrom exponent, single scattering albedo, phase functions, and asymmetry parameters) were computed using the AERONET inversion algorithm [32,33]. From the AERONET products, the monthly mean values of the AE at 380 nm and 500 nm and the AOD at 500 nm were selected for analysis. Standard deviation was used as a measure of variability of these parameters. For those stations where the 500-nm wavelength was not available, the AOD at 500 nm was calculated from the AOD at 440 nm and AE using equations from Soni et al. [34].

Annual and seasonal mean values of aerosol intensive optical parameters were calculated for the overall datasets or on each defined geographical cluster. Observing stations and the number of datasets considered in this study are presented in Table 1.

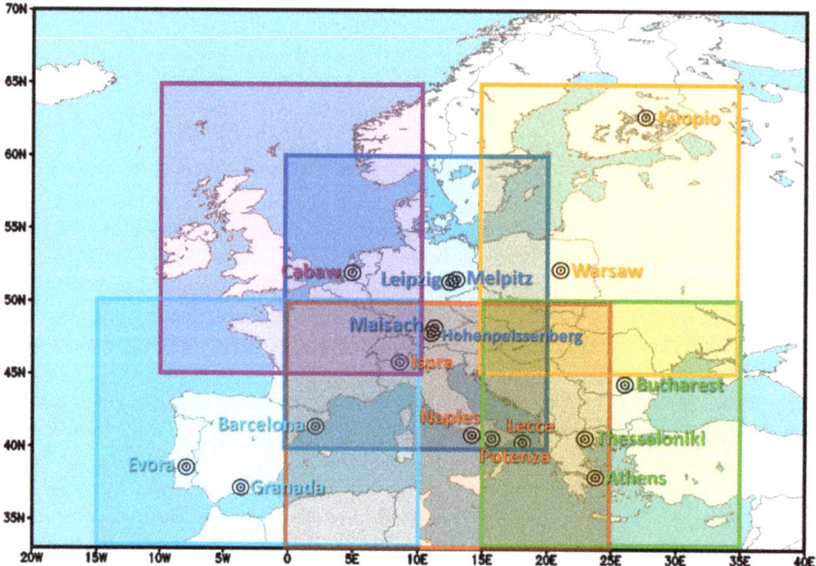

Figure 1. Cluster domains: Central (in blue, containing Hohenspeissenberg, Leipzig, Maisach, and Melpitz), Southeast (in green, containing Athens, Bucharest, and Thessaloniki), Mediterranean (in orange, containing Ispra, Lecce, Naples, and Potenza), Southwest (in cyan, containing Barcelona, Evora, and Granada), Northwest (in magenta, containing Cabaw), and Northeast (in yellow, containing Kuopio and Warsaw).

Table 1. List of observing stations and number of data points used (2008–2018). QA: quality assurance.

Station	Longitude (E)	Latitude (N)	Altitude (m a.s.l.)	Number of Data Points from Photometer (Monthly Means)	Number of Profiles from Lidar	Number of Layers from Lidar	Number of QA Layers from Lidar
Athens	23.78	37.96	212	128	63	302	37
Barcelona	2.12	41.393	115	132	2	9	2
Bucharest	26.029	44.348	93	132	227	880	126
Cabaw	4.93	51.97	0	132	37	213	26
Evora	−7.9115	38.5678	293	132	205	851	101
Granada	−3.605	37.164	680	132	120	679	106
Hohenpeissenberg	11.0119	47.8019	974	69	33	157	17
Ispra	8.6167	45.8167	209	132	17	121	10
Kuopio	27.55	62.7333	190	129	84	194	15
Lecce	18.1	40.333	30	132	16	66	0
Leipzig	12.433	51.35	90	132	62	284	62
Maisach	11.258	48.209	516	132	5	15	4
Melpitz	12.93	51.53	84	23	6	18	4
Naples	14.183	40.838	118	33	8	32	5
Potenza	15.72	40.6	720	132	137	702	153
Thessaloniki	22.95	40.63	50	132	70	202	111
Warsaw	20.98	52.21	112	11	37	232	29
Total				**1845**	**1129**	**4957**	**808**

2.2. Neural Network Aerosol Typing Algorithm Based on Lidar Data: NATALI Software

The aerosol typing of lidar data from the EARLINET database was done using the custom-made software NATALI (see Supplementary Materials). This classification algorithm has already been tested on synthetic and measurement data, and the entire processes of building up and software quality assurance are fully described in Nicolae D. et al. [20].

The NATALI software uses as input parameters the EARLINET database: the backscatter coefficient profiles at 1064, 532, and 355 nm; the extinction coefficient profiles at 532 and 355 nm; and optionally the linear particle depolarization profile at 532 nm. Depending on the available input parameters and their quality, the typing can be done at three levels: (1) high-resolution typing when depolarization is available (14 aerosol types (pure, mixtures of two, and mixtures of three pure aerosol types) can be identified); (2) low-resolution typing when depolarization is available (six predominant aerosol types (70% pure aerosols with traces from other types) can be identified) (used especially when the uncertainty of the input data is high); and (3) low-resolution typing when depolarization is not available (5 predominant aerosol types can be identified). In this article, the typing was performed using option 3 (low-resolution typing when depolarization is not available) because the calibrated depolarization profiles were included only recently in the EARLINET observations and therefore were not available for most of the years analyzed in this article. Dust, Smoke, Continental, Continental Polluted, and Marine/Cloud were the aerosol types retrieved using the low-resolution typing part of the NATALI software. Their principal characteristics and sources are in the table below (Table 2), and they have been described in Nicolae D. et al. [20]. Marine aerosols and aerosol layers corrupted by clouds cannot be separated by the algorithm when using the low-resolution typing level due to their similar optical properties at the lidar wavelengths.

Table 2. Conventional names of the aerosol types [20].

Aerosol Type	Source	Particle Characteristics
Continental	Land surfaces	Medium size, medium spherical, medium absorbing
Dust	Desert surfaces	Large, nonspherical, medium absorbing
Continental polluted	Industrial sites	Small, spherical, highly absorbing
Marine	Sea surface	Large, aspherical, nonabsorbing
Smoke	Vegetation fires	Small, spherical, highly absorbing

NATALI software detects the layer boundaries with the gradient method [35] using the 1064-nm backscatter coefficient profile. For each layer, the aerosol mean intensive optical parameters are calculated using the three backscatter coefficients and the two extinction coefficients, as follows:

$$\mathrm{AE}_{\lambda_1/\lambda_2} = -\frac{\ln \alpha_{\lambda_1}/\alpha_{\lambda_2}}{\ln \lambda_1/\lambda_2}, \tag{1}$$

$$\mathrm{CI}_{\lambda_1/\lambda_2} = -\frac{\ln \beta_{\lambda_1}/\beta_{\lambda_2}}{\ln \lambda_1/\lambda_2}, \tag{2}$$

$$\mathrm{LR}_{\lambda_1} = \frac{\alpha_{\lambda_1}}{\beta_{\lambda_1}}, \tag{3}$$

$$\mathrm{CR}_{\lambda_1/\lambda_2} = \frac{\beta_{\lambda_1}}{\beta_{\lambda_2}}, \tag{4}$$

where α represents the extinction coefficient, β the backscatter coefficient, and λ the wavelength.

The entire set of parameters (i.e., AEs, color ratios (CRs), color indexes (CIs), and LRs and PDRs when available) was used as input for several ANNs upon which the typing process was based. The ANNs were built on a synthetic database constructed from a customized aerosol model able to reproduce the necessary observed aerosol properties in a statistically relevant number. At least 3500 samples of each aerosol type were used as input to the ANNs for supervised learning. A cascade of three ANNs with different architectures was established to perform the classification on each typing level in order to maximize the performance of the typing process. The predominant aerosol type was selected using a voting procedure applied to the responses of the three ANNs based on the ANN output confidence level and stability of answers. The NATALI software performance in retrieving the aerosol type is highly dependent on the physical content of the optical inputs and their uncertainty and availability, and the predominant aerosol types that are recognized in more than 70% of the cases with intensive optical parameters have uncertainties lower than 50% [20].

2.3. Clusterization Based on Typical Atmospheric Circulation

To analyze the air masses carrying different aerosol types from various sources over Europe between 2008 and 2018, the GrossWetterTypes threshold-based catalogue with 18 circulation types (GWT18) was used [36]. The GWT classification catalogue was developed within COST 733 Action (the European Cooperation in Science and Technology) [37,38], and it organizes the maps into types according to varying degrees of zonality, meridionality, and vorticity determined as spatial correlation coefficients between daily sea level pressure fields and prototypical flow patterns [36].

The methodology proposed in COST 733 to identify circulation types was based on sea level pressure. In this article, instead of sea level pressure, a 850-hPa geopotential height was used, because this level is representative of the local circulation within the planetary boundary layer (PBL) and for long-range transport within the free troposphere. The geopotential height dataset was extracted from ECMWF ERA Interim Reanalysis (European Centre for Medium-Range Weather Forecasts) [39] at a $0.25° \times 0.25°$ horizontal resolution.

The analysis was performed in two steps. First, the circulation type around each station was assessed, and then the stations were grouped based on their common circulation features (Table 3). Around each cluster, an extended domain was defined (Figure 1). Second, the circulation types according to the GWT18 catalogue (not shown) were assessed for each cluster. The GWT18 catalogue discerned the dominant circulation over an area into cyclonic and anticyclonic types in eight cardinal and intercardinal directions (N, S, E, W, NE, SE, SW, NW) and discerned two undefined cyclonic and anticyclonic circulations.

Table 3. The clusters of European Aerosol Research Lidar Network (EARLINET) and Aerosol Robotic Network (AERONET) stations and their spatial extent.

Cluster	Spatial Extent (lon/lat)	Stations
Central	0°–20° E/40°–60° N	Hohenspeissenberg, Leipzig, Maisach, and Melpitz
Southeast	15°–35° E/30°–50° N	Athens, Bucharest, and Thessaloniki
Mediterranean	5°–25° E/30°–50° N	Ispra, Lecce, Naples, and Potenza
Southwest	15° W–10° E/30°–50° N	Barcelona, Evora, and Granada
Northwest	10° W–10° E/45°–65° N	Cabaw
Northeast	15°–35° E/45°–65° N	Kuopio and Warsaw

3. Results

3.1. Optical Characteristics of Aerosols over Europe

Mean values of the aerosols' optical parameters for Europe are presented in Table 4. These were obtained from EARLINET observations between 2008 and 2018. In total, 4957 layers were analyzed, and for each layer, the optical parameters (i.e., backscatter coefficients at 1064, 532, and 355 nm, extinction coefficients at 532 and 355 nm) were used to calculate the intensive optical parameters (Equations (1)–(3)). Table 4 shows the mean values obtained for all layers passing the QA criteria (808 layers) and their associated standard deviations. Based on the inversely proportional relationship between Angstrom exponents and the particle radius (e.g., References [40,41]), the results indicate that the typical aerosol particles observed during the study period over Europe were medium-sized absorbing particles with low spectral dependence (i.e., relatively similar values for the parameters in Table 4 [42]).

Table 4. Mean values of the intensive optical parameters over Europe between 2008 and 2018 as retrieved from EARLINET lidar measurements (808 layers that passed the QA criteria).

Parameter	Mean Value and Standard Deviation
Angstrom exponent (355/532 nm)	1.18 ± 0.14
Color index (355/532 nm)	1.31 ± 0.13
Color index (532/1064 nm)	0.93 ± 0.08
Lidar ratio (355 nm)	62 ± 3 sr
Lidar ratio (532 nm)	67 ± 4 sr

Differences between the optical properties of aerosols in different regions of Europe were analyzed. Figure 2 shows the mean aerosol intensive optical parameters for each geographical cluster, calculated from EARLINET lidar measurements and AERONET data. Although collocated, the two measurement techniques did not sample the same atmosphere because the Raman lidars are usually operated during nighttime, while the photometers are usually operated during daytime. This may introduce systematic biases when comparisons are conducted directly. In addition, lidar does not detect low-level aerosols in the first part of the planetary boundary layer (approximatively 200 up to 800 m) due to its technological limitations (the overlap issue [5]). This may also introduce biases as long as the locally produced aerosols are generally confined in the PBL and the lidar cannot measure their specific characteristics.

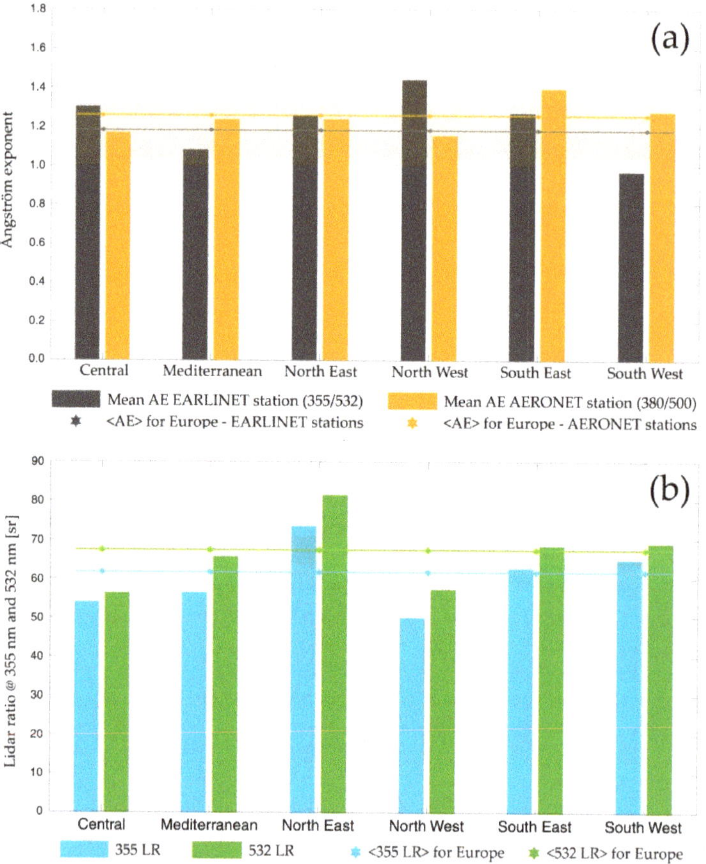

Figure 2. Optical properties of aerosols as retrieved from EARLINET lidar measurements and AERONET photometer measurements between 2008 and 2018: (**a**) mean values of Angstrom exponents from lidar (355/532 nm, dark gray bar) and from photometers (380/500 nm, orange bar) for each cluster; (**b**) mean values of lidar ratios from lidar at 355 nm (blue bar) and 532 nm (green bar) for each cluster. The mean European values (dark gray and orange lines for the Angstrom exponents and blue and green lines for the lidar ratios) and the standard deviations (shaded) are indicated.

The lowest mean values for the Angstrom exponent (i.e., 0.97) were observed over the Iberian Peninsula (Southwest cluster) and were associated with Dust and Marine particles (Figure 2a). The highest mean values for the lidar ratio at 532 nm (i.e., 82 sr) were recorded over Northeastern Europe and were associated with Smoke (Figure 2b).

Figure 3 shows the aerosol optical properties (i.e., AODs at 500 nm and Angstrom exponents at 380/500 nm) for each geographical cluster and season, calculated from AERONET photometer measurements. In total, 1845 monthly averaged values of AODs and AEs were used.

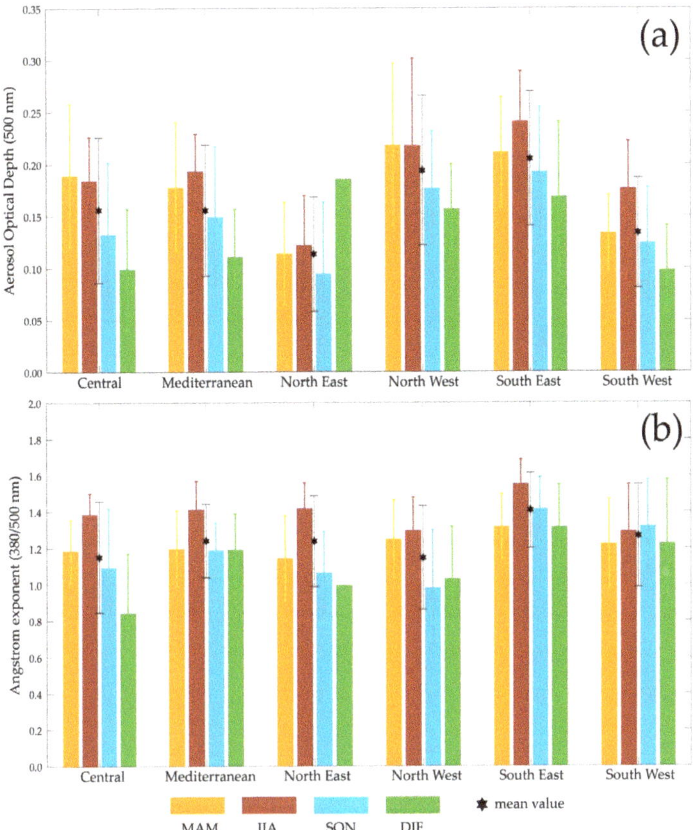

Figure 3. Optical properties of aerosols as retrieved from AERONET photometer measurements between 2008 and 2018: (**a**) aerosol optical depths (AODs) (500 nm) and (**b**) Angstrom exponents (AEs) (380/500 nm). The mean values and the standard deviations for the AODs and AEs for each cluster are indicated by a black marker and whiskers.

High AOD values (i.e., 0.3) were observed during summer over the Southeast (Figure 3a). The Northeast cluster was characterized by low AOD values for all seasons (with the exception of winter), which can be explained by cleaner air conditions compared to the other clusters (e.g., Reference [43]). The highest values for AE were observed during summer (i.e., months JJA (June, July, August)) and varied from 1.29 for the Southwest and Northwest clusters to 1.55 for the Southeast cluster. These high values can be explained by the presence of small particles (e.g., smoke) transported in the troposphere (Figure 3b). The smallest values for AE were recorded during winter (i.e., months DJF (December, January, February)) for all clusters, with the exception of the Northwest cluster, where the smallest AE values were recorded during autumn (i.e., months SON (September, October, November)) (Figure 3b) (e.g., Reference [13]).

3.2. Aerosol Types over Europe

The distribution of aerosol types in Europe shows that Smoke (37%) and Continental (25%) aerosol types were identified by NATALI software as the predominant aerosol types in Europe, followed by the Continental Polluted type (17%) and Marine/Cloud (10%). Dust (10%) is generally transported from the Sahara, but there is also a contribution from soil dust. Approximately 1% of the analyzed cases were not recognized by NATALI. Differences between aerosol types in different regions of Europe are shown in Figure 4.

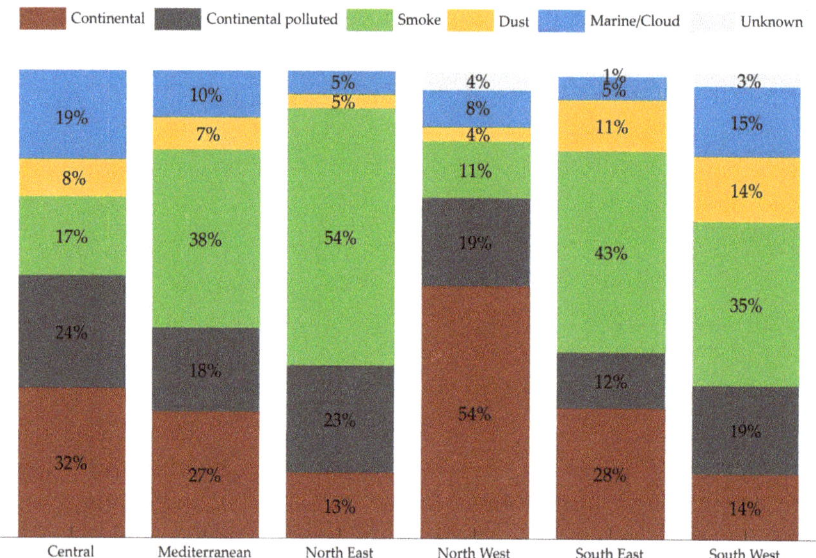

Figure 4. Prevalence of aerosol types retrieved with the Neural network Aerosol Typing Algorithm based on LIdar data (NATALI) code from EARLINET lidar measurements between 2008 and 2018.

The dominant aerosol type was Continental for the Northwest (54%) and Central (32%) clusters. The Marine/Cloud type was more frequently identified over Central Europe (19%) and the Southwest (15%), but this type was also present in the Mediterranean (10%) cluster. The high percentage obtained for Central Europe was very likely due to incomplete cloud screening, taking into account the drawback of the algorithm in the low-resolution typing level in separating the marine and cloud corrupted layers.

Over the Northeast (54%), Southeast (43%), and Mediterranean (38%), Smoke was the dominant aerosol type. This was most likely due to the forest fires that occur frequently over the Ukraine, the Russian Federation, Greece, and Turkey [44,45]. Smoke from Africa is also transported over the Iberian Peninsula and into the Mediterranean region [46].

In order to explain the aerosol types retrieved with NATALI code, for each domain, the GWT18 circulation types corresponding to the date of EARLINET measurements were analyzed (Figure 5). For the Central cluster, the dominant circulation type was western (34.9%), which enabled humid air mass advections from the Atlantic Ocean toward the Central domain, while the Southwestern circulation type (17.9%) enabled the transport of dry and hot air masses from North Africa. The western circulation explained the Marine/Cloud type and also part of the Smoke aerosols of North American origin [47]. The southwestern circulations enabled the Dust particle advection and also Smoke aerosols from Iberian Peninsula summer fires. The percentage of undefined anticyclonic circulation types (12.3%) was also remarkable, which is the signature of local circulation and could explain the high percentage of Continental and Continental Polluted aerosol types, as it is known that Central and Eastern Europe are powerful industrial areas known as "the black triangle" [48].

The northern, northwestern, and western anticyclonic circulation types over the Mediterranean cluster were quasi-equally represented (approx. 15% each), while northeastern circulations were 10.7% and southwestern circulations were 7.3%. The northern and northeastern circulation types were responsible for Continental and Continental Polluted aerosols, since the air masses cross "the black triangle". Western and northwestern circulations were responsible for long-range transport of

Smoke particles [49–51]. Southwestern circulations were mainly responsible for Dust particles and for Marine/Cloud aerosols.

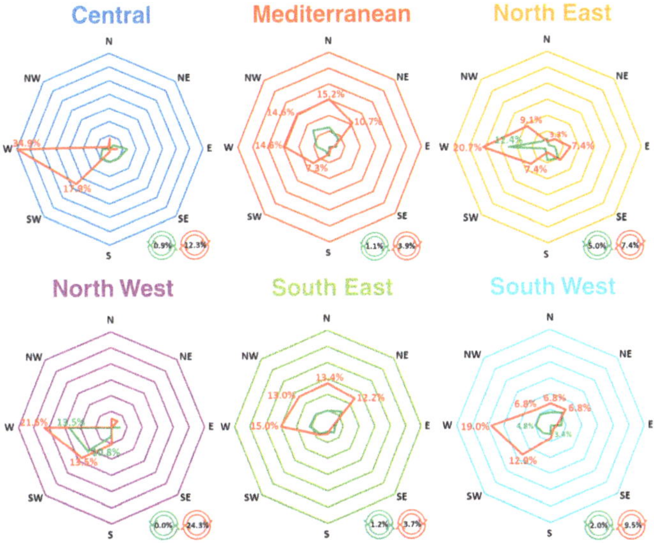

Figure 5. Circulation types based on the GWT18 circulation type catalogue corresponding to the clusters described in Figure 1 (color-coded accordingly with the colors for each cluster). Only data corresponding with the time of EARLINET lidar measurements were used. For each cluster, the wind direction is shown for cyclonic (red line) and anticyclonic (green line) circulations. The percentage shown for each cluster represents the total number of observations of the dominant wind direction. For each cluster, the percentage of undefined cyclonic (green circle) and anticyclonic (red circle) circulation is shown.

The dominant circulation type over the Northeastern cluster was western anticyclonic (20.7%) and cyclonic (12.4%), which explains the presence of Continental and Continental Polluted aerosols but also enabled the long-range transport of Smoke particles from North America [52,53] and from local fires [54]. Smoke aerosols from the Ukraine and the Russian Federation were also enabled by eastern (7.4%) and northeastern circulations (3.3%). The northwestern circulation type (9.1%) enabled the advection of Marine aerosols from the Baltic Sea, while southwestern circulation (7.4%) allowed the dry air masses to advect Dust particles from North Africa and Smoke particles from western Balkan fire events.

The highest percentage of circulation types over the Northwestern cluster was anticyclonic undefined (24.3%), which explains the highest percentage of Continental and Continental Polluted particles due to subsidence. The western circulation, both anticyclonic (21.6%) and cyclonic (13.5%), explained the trans-Atlantic transport of Smoke particles [55–57] and marine aerosols. The southwestern anticyclonic circulation (13.5%) was linked with Dust events [58] and Iberian fire events [59], while the cyclonic circulation (10.8%) explained the presence of Marine/Clouds.

The Southeastern cluster was characterized by western (15%), northern (13.4%), northwestern (13.0%), and northeastern (12.2%) anticyclonic circulation types. These circulation types were linked with Continental and Continental Polluted (western and northwestern circulations) and Smoke aerosols (northern and northeastern circulations). Southwestern circulations were mostly linked with Dust events, while southern, southeastern, and eastern circulation types were associated with Marine/Clouds.

Western (19.0% anticyclonic, 4.8% cyclonic) and southwestern (12.9% anticyclonic, 3.4% cyclonic) circulation types characterized the Southwestern cluster, explaining the presence of Dust particles [60] and Marine/Clouds. In addition, northwestern and northern and eastern circulations allowed the humid air masses to transport Marine/Clouds from the Atlantic and the Mediterranean Sea toward the Iberian Peninsula. The northeastern (6.8%) circulation type, together with the anticyclonic undefined (9.5%) circulations, explained the presence of Continental and Continental Polluted aerosols either from France or that were locally generated. Due to prolonged droughts and high temperatures during the warm seasons, numerous fire events were reported in Portugal and Spain (http://effis.jrc.ec.europa.eu/reports-and-publications/annual-fire-reports/ and Reference [59]).

Seasonal variation of aerosols over Europe is shown in Figure 6a. In this plot, the results are presented as percentages of the total number of aerosol layers observed during a certain season and between 2008 and 2018 to put into evidence the frequency of occurrence of different aerosol types. However, the number of layers varied with the season because of weather conditions, which made the sampling over the seasons inhomogeneous: a clear sky in summer (399 layers) and the presence of low clouds or precipitation in spring (203 layers), winter (72 layers), and autumn (134 layers). In addition, it should be noted that NATALI is not able to distinguish between Marine aerosols and aerosol layers corrupted by clouds. Cloud particles are large and low absorbing and are therefore very similar to Marine aerosols [20]. As such, lidar observations that are not completely cloud-free may result in a false classification as Marine. This is particularly true for the winter season, when the atmosphere is generally cloudy and lidar measurements are performed between broken clouds.

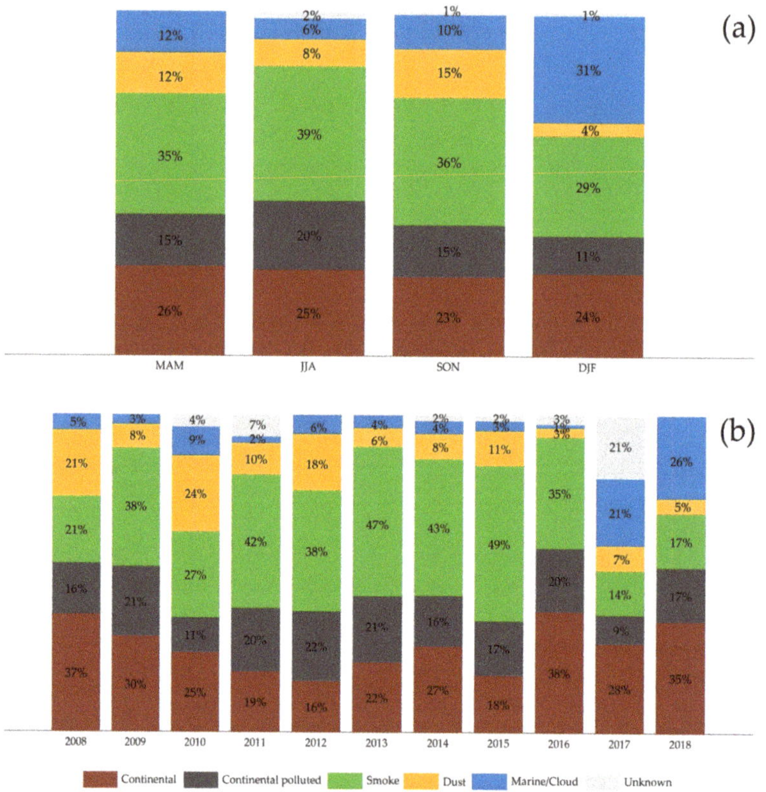

Figure 6. Time distribution of the prevalence of aerosol types retrieved with the NATALI code from EARLINET lidar measurements between 2008 and 2018: (**a**) seasonal distribution, (**b**) annual variations.

Except for winter (fewer data and potentially corrupted by clouds), the seasonal variability was insignificant at a continental scale: small increases in the percentage of Smoke during spring and small increases in Dust during autumn (out of the total number of layers measured in each season) (Figure 6a). The variation of aerosol types over Europe showed that the proportions of aerosol types varied each year, being influenced by long-range transport from other continents and by the meteorological situation.

Figure 6b shows that there was significant production and transport of biomass burning aerosols between 2011 and 2015 [4,47,61], while Dust transport increased in 2008, 2010, and 2012 [5,61]. The Dust category included, besides mineral dust, also volcanic dust due to the similar intensive optical parameters, which explains the increase in the percentage of Dust in 2010, when volcanic ash from the Eyjafjallajokull volcano was injected in the troposphere and transported over Europe [62]. In recent years (i.e., 2017–2018), more Marine/Clouds were identified by the NATALI algorithm. These data were not finalized in the EARLINET database, i.e., not all quality assurance procedures applied, and therefore biases due to an incomplete cloud screening may apply.

The years with a high percentage of Smoke in Figure 6b generally corresponded with the years with high temperatures and prolonged droughts, which were the key drivers of warm season fires across the European region [63]. Since the beginning of the 21st century, Europe has experienced a series of extreme hot and dry summers [64] covering different parts of the continent. The years of 2014 (43%) and 2015 (49%) were among the top 10 European record-breaking heatwave events since 1950 [65,66]. Heat waves episodes were recorded almost every summer in different European regions. According to the Joint Research Centre's annual reports on "Forest Fires in Europe, the Middle East, and North Africa" (http://effis.jrc.ec.europa.eu/reports-and-publications/annual-fire-reports/), the most affected European area was the Mediterranean region (Portugal, Spain, France, Italy, and Greece). The highest peaks of fire activities for this region were registered during the entire analyzed period excepting 2010 and 2014. In addition, fire events were recorded in western, central, and northern Europe during 2011 and 2013–2018. Smoke particles of North American forest fire origin were reported over the Mediterranean Basin in 2013 [49,50] and 2014 [51], over western Europe in 2017 [56], over northern Europe in 2016 [55], and over central Europe in 2013 [47,52] and 2015 [53].

The annual variability of the optical properties of aerosols over Europe is presented in Figure 7, as obtained from AERONET measurements. The European mean values are highlighted in the time evolution of the AOD and AE for all clusters (including the standard deviations as shaded areas) using linear and fourth-grade polynomial fits, respectively (black lines). The decreasing trend of the AOD is clearly visible in these results, and the linear fit of the European mean values had a correlation coefficient of 0.62. The AOD trend pointed to an overall decrease in the aerosol concentration in Europe, probably due to the implementation of environmental policies [67–69]. However, large values of AODs were measured in the Southeast in 2008 and 2011 [3], as well as in the Northwest in 2015 [70]. In these regions, the AOD was always above the European mean, pointing to higher concentrations of aerosols but of different types. In the Southeast, the high AODs were associated with persistent Smoke transport in the troposphere from the east and south, while in the Northwest, the high AODs were more related to the transport of Continental Polluted aerosol from Central Europe on top of locally produced particles. The Northeast and Southwest were constantly the "cleanest" regions in Europe, i.e., having the smallest AOD values recorded by AERONET.

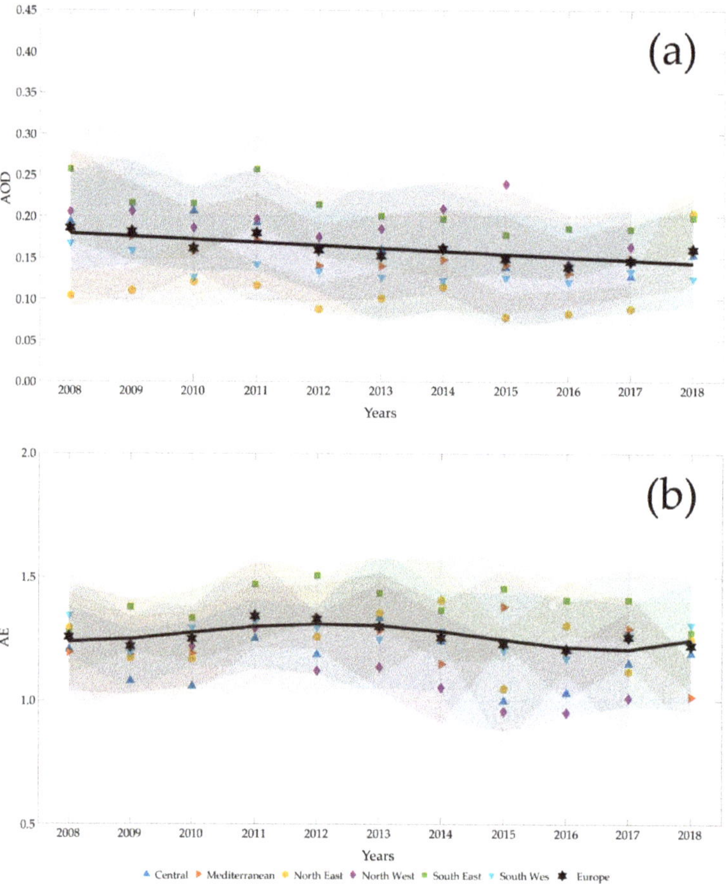

Figure 7. Changes in the aerosol optical properties over Europe between 2008 and 2018 as measured by AERONET photometers: (**a**) time evolution of the aerosol optical depths (500 nm); (**b**) time evolution of the Angstrom exponents (380/500 nm). Black lines represent the linear and fourth-grade polynomial fit of the European mean values, and the shaded areas represent the standard deviations.

Aerosol AEs varied over Europe with values greater than 1.2 between 2011 and 2014 and values lower than 1.2 for the rest of the study period. The high values of AEs were associated with an increased proportion of Smoke over Europe (Figure 7b). The largest value for the AEs (i.e., 1.5) was recorded for the Southeast cluster in 2012, very likely associated with the transport of biomass burning aerosols over Bucharest and Athens. An interesting feature is the change in the AE in the Northwest region, from values around the European mean between 2008 and 2011 to very low values starting in 2012. The Southeast and Mediterranean regions seemed to have the same phase of oscillations for the AE, opposite to Central Europe. The years 2012 and 2015 were characterized by high AEs (small particles) in the southeast and Mediterranean clusters, but low AEs (larger particles) in central Europe. On the contrary, in 2013, the AE decreased in the Southeast and Mediterranean regions and increased in Central Europe.

In general, the differences between regions in terms of AE values (and therefore the size of the particles) increased with time. The values of the Angstrom exponent varied significantly from year to year, showing an important contribution from the long-range transported aerosols to the total aerosol load in the column.

4. Conclusions

In this study, aerosols over Europe were characterized based on lidar and photometer observations collected between 2008 and 2018 by the EARLINET (European Aerosol Research Lidar Network) and AERONET (Aerosol Robotic Network). Seventeen observing stations distributed over six regions in Europe and operating both multiwavelength Raman lidars and automatic sun/sky photometers were selected.

In order to retrieve the aerosol type, NATALI (Neural Network Aerosol Typing Algorithm Based on Lidar Data) software was used. NATALI identifies the aerosol layer boundaries with the gradient method applied on the backscatter coefficient at 1064 nm, calculates the mean layer aerosol intensive optical parameters, and runs a combination of artificial neural networks to identify the most probable aerosol type that fits the optical data.

The analysis showed that the typical aerosol particle observed over Europe between 2008 and 2018 was of medium size, had medium absorbing particles, and had low spectral dependence. The aerosol properties had a large variability in time (from season to season and year to year) compared to the variability in space (from region to region) and were largely influenced by long-range transport from other continents and by large-scale circulation.

Smoke was more present in the East (55% out of all layers observed in the North and 43% in the South) but could also be found in the Mediterranean region (38%) and over the Iberian Peninsula (35%) due to frequent transport from Africa. Central and Northwestern Europe were generally dominated by the Continental aerosols (32% and 54%, respectively). Dust (locally produced soil dust and long-range transported desert dust) was measured all over Europe, as already highlighted by several papers, but mostly in autumn (15%) and spring (12%), when it is lifted by the turbulent air masses above the PBL. Continental Polluted aerosols represented a high proportion (17%) of all aerosol types in Europe, although the decreasing AOD may have been partially related to the implementation of European environmental regulations.

Years with high temperatures in Europe (2011–2015) were highlighted also by an increase in the Angstrom exponent measured by AERONET (>1.3) as well as by an increase in the proportion of Smoke particles derived by NATALI from EARLINET observations (more than 35% out of the total number of aerosol layers).

Based on the decadal variation of the Angstrom exponents, these "hot and smoky" years (2011–2015) seemed to be balanced by "dusty" years (2008–2010) and "wet" years (2017–2018); however, they were not equal in all parts of Europe. The high variability from year to year showed that aerosol transport in the troposphere became more and more important in the overall balance of the columnar aerosol load and that regional characteristics were no longer linked to close-by aerosol sources but to dominant circulations bringing dust and biomass burning aerosols from other continents.

Supplementary Materials: The NATALI (Neural Network Aerosol Typing Algorithm Based on Lidar Data) software is available with a user guide from http://natali.inoe.ro/resources.html/software.

Author Contributions: Conceptualization, D.N.; methodology, S.A., D.E., and J.V.; software, V.N. and T.S.; formal analysis, C.T. and A.D.; investigation, S.A. and B.A.; writing—original draft preparation, B.A. and V.N; writing—all authors; writing—review and editing, B.A and J.V.; visualization, D.E. and V.-E.T.; supervision, D.N. and S.S.

Funding: This research was funded by the Ministry of Research and Innovation through Program I—Development of the national research development system, Subprogram 1.2—Institutional Performance—Projects of Excellence Financing in RDI, Contract No.19PFE/17.10.2018, by the Romanian National Core Program Contract No.18N/2019; by the RDI Program for Space Technology and Advanced Research contract 183/2017; and by the European Regional Development Fund through the Competitiveness Operational Programme 2014–2020, POC-A.1-A.1.1.1-F-2015, project Research Centre for environment and Earth Observation CEO-Terra, SMIS code 108109, contract No. 152/2016. Camelia Talianu was supported by the Austrian Science Fund (FWF), Project M 2031, Meitner-Programm.

Acknowledgments: The authors acknowledge EARLINET for providing aerosol lidar profiles available from the EARLINET webpage. We would like to acknowledge the AERONET team for calibrating and processing the data and to thank the principal investigators and their staff for maintaining the 17 sites identified in Section 2.1.

Conflicts of Interest: The authors declare no conflict of interest.

References

1. Wang, H.; Xie, X.; Liu, X. On the Robustness of the Weakening Effect of Anthropogenic Aerosols on the East Asian Summer Monsoon with Multimodel Results. *Adv. Meteorol.* **2015**, *2015*, 397395. [CrossRef]
2. Farahat, A. Comparative analysis of MODIS, MISR, and AERONET climatology over the Middle East and North Africa. *Ann. Geophys.* **2019**, *37*, 49–64. [CrossRef]
3. Siomos, N.; Balis, D.S.; Voudouri, K.A.; Giannakaki, E.; Filioglou, M.; Amiridis, V.; Papayannis, A.; Fragkos, K. Are EARLINET and AERONET climatologies consistent? The case of Thessaloniki, Greece. *Atmos. Chem. Phys.* **2018**, *18*, 11885–11903. [CrossRef]
4. Georgoulias, A.K.; Alexandri, G.; Kourtidis, K.A.; Lelieveld, J.; Zanis, P.; Pöschl, U.; Levy, R.; Amiridis, V.; Marinou, E.; Tsikerdekis, A. Spatiotemporal variability and contribution of different aerosol types to the aerosol optical depth over the Eastern Mediterranean. *Atmos. Chem. Phys.* **2016**, *16*, 13853–13884. [CrossRef] [PubMed]
5. Wandinger, U.; Freudenthaler, V.; Baars, H.; Amodeo, A.; Engelmann, R.; Mattis, I.; Groß, S.; Pappalardo, G.; Giunta, A.; D'Amico, G.; et al. EARLINET instrument intercomparison campaigns: Overview on strategy and results. *Atmos. Meas. Tech.* **2016**, *9*, 1001–1023. [CrossRef]
6. Pappalardo, G.; Amodeo, A.; Apituley, A.; Comeron, A.; Freudenthaler, V.; Linné, H.; Ansmann, A.; Bösenberg, J.; D'Amico, G.; Mattis, I.; et al. EARLINET: Towards an advanced sustainable European aerosol lidar network. *Atmos. Meas. Tech.* **2014**, *7*, 2389–2409. [CrossRef]
7. Burton, S.P.; Ferrare, R.A.; Vaughan, M.A.; Omar, A.H.; Rogers, R.R.; Hostetler, C.A.; Hair, J.W. Aerosol classification from airborne HSRL and comparisons with the CALIPSO vertical feature mask. *Atmos. Meas. Tech.* **2013**, *6*, 1397–1412. [CrossRef]
8. Groß, S.; Esselborn, M.; Weinzierl, B.; Wirth, M.; Fix, A.; Petzold, A. Aerosol classification by airborne high spectral resolution lidar observations. *Atmos. Chem. Phys.* **2013**, *13*, 2487–2505. [CrossRef]
9. Papagiannopoulos, N.; Mona, L.; Amodeo, A.; D'Amico, G.; Gumà Claramunt, P.; Pappalardo, G.; Alados-Arboledas, L.; Guerrero-Rascado, J.L.; Amiridis, V.; Kokkalis, P.; et al. An automatic observation-based aerosol typing method for EARLINET. *Atmos. Chem. Phys.* **2018**, *18*, 15879–15901. [CrossRef]
10. Wandinger, U.; Baars, H.; Engelmann, R.; Hünerbein, A.; Horn, S.; Kanitz, T.; Donovan, D.; van Zadelhoff, G.-J.; Daou, D.; Fischer, J.; et al. HETEAC: The Aerosol Classification Model for EarthCARE. *EPJ Web Conf.* **2016**, *119*, 01004. [CrossRef]
11. Holben, B.N.; Eck, T.F.; Slutsker, I.; Tanre, D.; Buis, J.P.; Setzer, A.; Vermote, E.; Reagan, J.A.; Kaufman, Y.; Nakajima, T.; et al. AERONET—A federated instrument network and data archive for aerosol characterization. *Remote Sens. Environ.* **1998**, *66*, 1–16. [CrossRef]
12. Giles, D.M.; Sinyuk, A.; Sorokin, M.G.; Schafer, J.S.; Smirnov, A.; Slutsker, I.; Eck, T.F.; Holben, B.N.; Lewis, J.R.; Campbell, J.R.; et al. Advancements in the Aerosol Robotic Network (AERONET) Version 3 database – automated near-real-time quality control algorithm with improved cloud screening for Sun photometer aerosol optical depth (AOD) measurements. *Atmos. Meas. Tech.* **2019**, *12*, 169–209. [CrossRef]
13. Giles, D.M.; Holben, B.N.; Eck, T.F.; Sinyuk, A.; Smirnov, A.; Slutsker, I.; Dickerson, R.; Thompson, A.; Schafer, J. An analysis of AERONET aerosol absorption properties and classifications representative of aerosol source regions. *J. Geophys. Res. Atmos.* **2012**. [CrossRef]
14. Shin, S.-K.; Tesche, M.; Noh, Y.; Müller, D. Aerosol-type classification based on AERONET version 3 inversion products. *Atmos. Meas. Tech. Discuss.* **2019**, *12*, 3789–3803. [CrossRef]
15. Tan, F.; Lim, H.S.; Abdullah, K.; Yoon, T.L.; Holben, B. AERONET data–based determination of aerosol types. *Atmos. Pollut. Res.* **2015**, *6*, 682–695. [CrossRef]
16. Russell, P.B.; Bergstrom, R.W.; Shinozuka, Y.; Clarke, A.D.; DeCarlo, P.F.; Jimenez, J.L.; Livingston, J.M.; Redemann, J.; Dubovik, O.; Strawa, A. Absorption Angstrom Exponent in AERONET and related data as an indicator of aerosol composition. *Atmos. Chem. Phys.* **2010**, *10*, 1155–1169. [CrossRef]
17. Eck, T.F.; Holben, B.N.; Reid, J.S.; Sinyuk, A.; Hyer, E.J.; O'Neill, N.T.; Shaw, G.E.; Vande Castle, J.R.; Chapin, F.S.; Dubovik, O.; et al. Optical properties of boreal region biomass burning aerosols in central Alaska and seasonal variation of aerosol optical depth at an Arctic coastal site. *J. Geophys. Res.* **2009**, *114*, D11201. [CrossRef]

18. Nemuc, A.; Vasilescu, J.; Talianu, C.; Belegante, L.; Nicolae, D. Assessment of aerosol's mass concentrations from measured linear particle depolarization ratio (vertically resolved) and simulations. *Atmos. Meas. Tech.* **2013**, *6*, 3243–3255. [CrossRef]

19. Groß, S.; Freudenthaler, V.; Wirth, M.; Weinzierl, B. Towards an aerosol classification scheme for future EarthCARE lidar observations and implications for research needs. *Atmos. Sci. Lett.* **2015**, *16*, 77–82. [CrossRef]

20. Nicolae, D.; Vasilescu, J.; Talianu, C.; Binietoglou, I.; Nicolae, V.; Andrei, S.; Antonescu, B. A neural network aerosol-typing algorithm based on lidar data. *Atmos. Chem. Phys.* **2018**, *18*, 14511–14537. [CrossRef]

21. Mamun, M.M.; Müller, D. Retrieval of Intensive Aerosol Microphysical Parameters from Multiwavelength Raman/HSRL Lidar: Feasibility Study with Artificial Neural Networks. *Atmos. Meas. Tech. Discuss.* **2016**. [CrossRef]

22. Lanzaco, B.L.; Olcese, L.E.; Palancar, G.G.; Toselli, B.M. An Improved Aerosol Optical Depth Map Based on Machine-Learning and MODIS Data: Development and Application in South America. *Aerosol. Air. Qual. Res.* **2017**, *17*, 1623–1636. [CrossRef]

23. Gangwar, R.K.; Mathur, A.K.; Gohil, B.S.; Basu, S. Neural Network Based Retrieval of Atmospheric Temperature Profile Using AMSU-A Observations. *Int. J. Atmos. Sci.* **2014**, *2014*, 763060. [CrossRef]

24. EARLINET. EARLINET Data Portal. Available online: https://data.earlinet.org/earlinet/login.zul;jsessionid=AA323AC6D1261F5B01BF3203A131CA84 (accessed on 1 April 2019).

25. AERONET. Available online: https://aeronet.gsfc.nasa.gov/cgi-bin/draw_map_display_aod_v3 (accessed on 1 April 2019).

26. Ansmann, A.; Riebesell, M.; Wandinger, U.; Weitkamp, C.; Voss, E.; Lahmann, W.; Michaelis, W. Combined raman elastic-backscatter LIDAR for vertical profiling of moisture, aerosol extinction, backscatter, and LIDAR ratio. *Appl. Phys. B* **1992**, *55*, 18–28. [CrossRef]

27. Fernald, F.G. Analysis of atmospheric lidar observations: Some comments. *Appl. Opt.* **1984**, *23*, 652–653. [CrossRef] [PubMed]

28. Klett, J.D. Stable analytical inversion solution for processing lidar returns. *Appl. Opt.* **1981**, *20*, 211–220. [CrossRef]

29. Freudenthaler, V. About the effects of polarising optics on lidar signals and the Δ90 calibration. *Atmos. Meas. Tech.* **2016**, *9*, 4181–4255. [CrossRef]

30. Belegante, L.; Bravo-Aranda, J.A.; Freudenthaler, V.; Nicolae, D.; Nemuc, A.; Ene, D.; Alados-Arboledas, L.; Amodeo, A.; Pappalardo, G.; D'Amico, G.; et al. Experimental techniques for the calibration of lidar depolarization channels in EARLINET. *Atmos. Meas. Tech.* **2018**, *11*, 1119–1141. [CrossRef]

31. D'Amico, G.; Amodeo, A.; Baars, H.; Binietoglou, I.; Freudenthaler, V.; Mattis, I.; Wandinger, U.; Pappalardo, G. EARLINET Single Calculus Chain—Overview on methodology and strategy. *Atmos. Meas. Tech.* **2015**, *8*, 4891–4916.

32. Dubovik, O.; King, M.D. A flexible inversion algorithm for retrieval of aerosol optical properties from Sun and sky radiance measurements. *J. Geophys. Res.* **2000**, *105*, 20673–20696. [CrossRef]

33. Dubovik, O.; Holben, B.N.; Eck, T.F.; Smirnov, A.; Kaufman, Y.J.; King, M.D.; Tanre, D.; Slutsker, I. Variability of absorption and optical properties of key aerosol types observed in worldwide locations. *J. Atmos. Sci.* **2002**, *59*, 590–608. [CrossRef]

34. Soni, K.; Singh, S.; Bano, T.; Tanwar, R.S.; Nath, S. Wavelength Dependence of the Aerosol Angstrom Exponent and Its Implications Over Delhi, India. *Aerosol. Sci. Tech.* **2011**, *45*, 1488–1498. [CrossRef]

35. Belegante, L.; Nicolae, D.; Nemuc, A.; Talianu, C.; Derognat, C. Retrieval of the boundary layer height from active and passive remote sensors. Comparison with a NWP model. *Acta Geophys.* **2014**, *62*, 276–289. [CrossRef]

36. Beck, C.; Jacobeit, J.; Jones, P.D. Frequency and within-type variations of large-scale circulation types and their effects on low-frequency climate variability in Central Europe since 1780. *Int. J. Climatol.* **2007**, *27*, 473–491. [CrossRef]

37. Tveito, O.E.; Huth, R. Circulation-type classifications in Europe: Results of the COST 733 Action. *Int. J. Climatol.* **2016**, *36*, 2671–2672. [CrossRef]

38. Philipp, A.; Beck, C.; Huth, R.; Jacobeit, J. Development and comparison of circulation type classifications using the COST 733 dataset and software. *Int. J. Climatol.* **2016**, *36*, 2673–2691. [CrossRef]

39. Dee, D.P.; Uppala, S.M.; Simmons, A.J.; Berrisford, P.; Poli, P.; Kobayashi, S.; Andrae, U.; Balmaseda, M.A.; Balsamo, G.; Bauer, P.; et al. The ERA-Interim reanalysis: Configuration and performance of the data assimilation system. *Q. J. Roy. Meteor. Soc.* **2011**, *137*, 553–597. [CrossRef]

40. Muller, D.; Mattis, I.; Ansmann, A.; Wandinger, U.; Ritter, C.; Kaiser, D. Multiwavelength Raman lidar observations of particle growth during long-range transport of forest-fire smoke in the free troposphere. *Geophys. Res. Lett.* **2007**, *34*, L05803. [CrossRef]

41. Schuster, G.L.; Dubovik, O.; Holben, B.N. Angstrom exponent and bimodal aerosol size distributions. *J. Geophys. Res.* **2006**, *111*, D07207. [CrossRef]

42. Cao, N.; Yang, S.; Cao, S.; Yang, S.; Shen, J. Accuracy calculation for lidar ratio and aerosol size distribution by dual-wavelength lidar. *Appl. Phys. A* **2019**, *125*, 590. [CrossRef]

43. Chubarova, N.Y.; Poliukhov, A.A.; Gorlova, I.D. Long-term variability of aerosol optical thickness in Eastern Europe over 2001–2014 according to the measurements at the Moscow MSU MO AERONET site with additional cloud and NO2 correction. *Atmos. Meas. Tech.* **2016**, *9*, 313–334. [CrossRef]

44. Nicolae, D.; Nemuc, A.; Müller, D.; Talianu, C.; Vasilescu, J.; Belegante, L.; Kolgotin, A. Characterization of fresh and aged biomass burning events using multiwavelength Raman lidar and mass spectrometry. *J. Geophys. Res.* **2013**, *118*, 2956–2965. [CrossRef]

45. Galytska, E.; Danylevsky, V.; Hommel, R.; Burrows, J.P. Increased aerosol content in the atmosphere over Ukraine during summer 2010. *Atmos. Meas. Tech.* **2018**, *11*, 2101–2118. [CrossRef]

46. Cachorro, V.E.; Toledano, C.; Prats, N.; Sorribas, M.; Mogo, S.; Berjón, A.; Torres, B.; Rodrigo, R.; de la Rosa, J.; DeFrutos, A.M. The strongest desert dust intrusion mixed with smoke over the Iberian Peninsula registered with Sun photometry. *J. Geophys. Res.* **2008**, *113*, D14S04. [CrossRef]

47. Ortiz-Amezcua, P.; Guerrero-Rascado, J.L.; Granados-Muñoz, M.J.; Benavent-Oltra, J.A.; Böckmann, C.; Samaras, S.; Stachlewska, I.S.; Janicka, Ł.; Baars, H.; Bohlmann, S.; et al. Microphysical characterization of long-range transported biomass burning particles from North America at three EARLINET stations. *Atmos. Chem. Phys.* **2017**, *17*, 5931–5946. [CrossRef]

48. Suchara, I.; Sucharova, J. Distribution of Sulphur and Heavy Metals in Forest Floor Humus of the Czech Republic. *Water. Air Soil Pollut.* **2002**, *136*, 289–316. [CrossRef]

49. Ancellet, G.; Pelon, J.; Totems, J.; Chazette, P.; Bazureau, A.; Sicard, M.; Di Iorio, T.; Dulac, F.; Mallet, M. Long-range transport and mixing of aerosol sources during the 2013 North American biomass burning episode: Analysis of multiple lidar observations in the western Mediterranean basin. *Atmos. Chem. Phys.* **2016**, *16*, 4725–4742. [CrossRef]

50. Chazette, P.; Totems, J.; Ancellet, G.; Pelon, J.; Sicard, M. Temporal consistency of lidar observations during aerosol transport events in the framework of the ChArMEx/ADRIMED campaign at Minorca in June 2013. *Atmos. Chem. Phys.* **2016**, *16*, 2863–2875. [CrossRef]

51. Brocchi, V.; Krysztofiak, G.; Catoire, V.; Guth, J.; Marécal, V.; Zbinden, R.; El Amraoui, L.; Dulac, F.; Ricaud, P. Intercontinental transport of biomass burning pollutants over the Mediterranean Basin during the summer 2014 ChArMEx-GLAM airborne campaign. *Atmos. Chem. Phys.* **2018**, *18*, 6887–6906. [CrossRef]

52. Janika, L.; Stachlewska, I.S.; Veselovskii, I.; Baars, H. Temporal variations in optical and microphysical properties of mineral dust and biomass burning aerosol derived from daytime Raman lidar observations over Warsaw, Poland. *Atmos. Environ.* **2017**, *169*, 162–174. [CrossRef]

53. Janicka, L.; Stachlewska, I.S. Properties of biomass burning aerosol mixtures derived at fine temporal and spatial scales from Raman lidar measurements: Part I optical properties. *Atmos. Chem. Phys. Discuss.* **2019**. in review. [CrossRef]

54. Martinsson, J.; Monteil, G.; Sporre, M.K.; Kaldal Hansen, A.M.; Kristensson, A.; Eriksson Stenström, K.; Swietlicki, E.; Glasius, M. Exploring sources of biogenic secondary organic aerosol compounds using chemical analysis and the FLEXPART model. *Atmos. Chem. Phys.* **2017**, *17*, 11025–11040. [CrossRef]

55. Vaughan, G.; Draude, A.P.; Ricketts, H.M.A.; Schultz, D.M.; Adam, M.; Sugier, J.; Wareing, D.P. Transport of Canadian forest fire smoke over the UK as observed by lidar. *Atmos. Chem. Phys.* **2018**, *18*, 11375–11388. [CrossRef]

56. Hu, Q.; Goloub, P.; Veselovskii, I.; Bravo-Aranda, J.-A.; Popovici, I.E.; Podvin, T.; Haeffelin, M.; Lopatin, A.; Dubovik, O.; Pietras, C.; et al. Long-range-transported Canadian smoke plumes in the lower stratosphere over northern France. *Atmos. Chem. Phys.* **2019**, *19*, 1173–1193.

57. Talianu, C.; Seibert, P. Analysis of sulfate aerosols over Austria: A case study. *Atmos. Chem. Phys.* **2019**, *19*, 6235–6250. [CrossRef]

58. Vieno, M.; Heal, M.R.; Twigg, M.M.; MacKenzie, I.A.; Braban, C.F.; Lingard, J.J.N.; Ritchie, S.; Beck, R.C.; Moring, A.; Ots, R.; et al. The UK particulate matter air pollution episode of March–April 2014: More than Saharan dust. *Environ. Res. Lett.* **2016**, *11*, 044004. [CrossRef]

59. Nunes, L.; Álvarez-González, J.; Alberdi, I.; Silva, V.; Rocha, V.; Castro Rego, F. Analysis of the occurrence of wildfires in the Iberian Peninsula based on harmonised data from national forest inventories. *Ann. For. Sci.* **2019**, *76*, 27. [CrossRef]

60. Cachorro, V.E.; Burgos, M.A.; Mateos, D.; Toledano, C.; Bennouna, Y.; Torres, B.; de Frutos, Á.M.; Herguedas, Á. Inventory of African desert dust events in the north-central Iberian Peninsula in 2003–2014 based on sun-photometer–AERONET and particulate-mass–EMEP data. *Atmos. Chem. Phys.* **2016**, *16*, 8227–8248. [CrossRef]

61. Granados-Muñoz, M.J.; Navas-Guzmán, F.; Guerrero-Rascado, J.L.; Bravo-Aranda, J.A.; Binietoglou, I.; Pereira, S.N.; Basart, S.; Baldasano, J.M.; Belegante, L.; Chaikovsky, A.; et al. Profiling of aerosol microphysical properties at several EARLINET/AERONET sites during the July 2012 ChArMEx/EMEP campaign. *Atmos. Chem. Phys.* **2016**, *16*, 7043–7066. [CrossRef]

62. Mortier, A.; Goloub, P.; Podvin, T.; Deroo, C.; Chaikovsky, A.; Ajtai, N.; Blarel, L.; Tanre, D.; Derimian, Y. Detection and characterization of volcanic ash plumes over Lille during the Eyjafjallajökull eruption. *Atmos. Chem. Phys.* **2013**, *13*, 3705–3720. [CrossRef]

63. Turco, M.; von Hardenberg, J.; AghaKouchak, A.; Llasat, M.C.; Provenzale, A.; Trigo, R. On the key role of droughts in the dynamics of summer fires in Mediterranean Europe. *Sci. Rep.* **2017**, *7*, 81. [CrossRef] [PubMed]

64. Hanel, M.; Rakovec, O.; Markonis, Y.; Máca, P.; Samaniego, L.; Kyselý, J.; Kumar, R. Revisiting the recent European droughts from a long-term perspective. *Sci. Rep.* **2018**, *8*, 9499. [CrossRef] [PubMed]

65. Russo, S.; Sillmann, J.; Fischer, E. Top ten European heatwaves since 1950 and their occurrence in the coming decades. *Environ. Res. Lett.* **2015**, *10*, 124003. [CrossRef]

66. Ionita, M.; Tallaksen, L.M.; Kingston, D.G.; Stagge, J.H.; Laaha, G.; van Lanen, H.A.J.; Scholz, P.; Chelcea, S.M.; Haslinger, K. The European 2015 drought from a climatological perspective. *Hydrol. Earth Syst. Sci.* **2017**, *21*, 1397–1419. [CrossRef]

67. DIRECTIVE (EU) 2016/2284 OF THE EUROPEAN PARLIAMENT AND OF THE COUNCIL on the reduction of national emissions of certain atmospheric pollutants, amending Directive 2003/35/EC and repealing Directive 2001/81/EC. Available online: https://eur-lex.europa.eu/legal-content/EN/ALL/?uri=CELEX%3A32016L2284 (accessed on 9 August 2019).

68. DIRECTIVE (EU) 2015/2193 OF THE EUROPEAN PARLIAMENT AND OF THE COUNCIL on the limitation of emissions of certain pollutants into the air from medium combustion plants. Available online: https://eur-lex.europa.eu/legal-content/EN/TXT/?uri=CELEX%3A32015L2193 (accessed on 9 August 2019).

69. Provençal, S.; Pavel Kishcha, P.; da Silva, A.M.; Elhacham, E.; Alpert, P. AOD distributions and trends of major aerosol species over a selection of the world's most populated cities based on the 1st Version of NASA's MERRA Aerosol Reanalysis. *Urban. Clim.* **2017**, *20*, 168–191. [CrossRef]

70. Markowicz, K.; Pakszys, P.; Ritter, C.; Zielinski, T.; Udisti, R.; Cappelletti, D.; Mazzola, M.; Shiobara, M.; Lynch, P.; Zawadzka, O.; et al. Impact of North American intense fires on aerosol optical properties measured over the European Arctic in July 2015. *J. Geophys. Res. Atmos.* **2016**, *121*, 14487–14512. [CrossRef]

Article

Wintertime Variations of Gaseous Atmospheric Constituents in Bucharest Peri-Urban Area

Cristina Antonia Marin [1,2], Luminiţa Mărmureanu [1,*], Cristian Radu [1], Alexandru Dandocsi [1,2], Cristina Stan [2], Flori Ţoancă [1], Liliana Preda [2] and Bogdan Antonescu [1,3]

[1] National Institute of Research and Development for Optoelectronics INOE 2000, Str. Atomiştilor 409, Măgurele, RO077125 Ilfov, Romania
[2] Department of Physics, Politehnica University of Bucharest, Str. Spl. Independenţei 313, RO060042 Bucureşti, Romania
[3] Faculty of Physics, University of Bucharest, RO077125 Bucureşti, Romania
[*] Correspondence: mluminita@inoe.ro

Received: 11 July 2019; Accepted: 16 August 2019; Published: 20 August 2019

Abstract: An intensive winter campaign was organized for measuring the surface air pollutants in southeastern Europe. For a three months period, the gas concentrations of NO_x, SO_2, CO, O_3, and CH_4 as well as meteorological parameters were simultaneously sampled to evaluate the variations and characteristic reactions between the gases during winter at the measuring site. The photochemical production of the ozone was observed through the diurnal variation of ozone and the solar radiation, the maximum concentration for ozone being reached one hour after the maximum value for solar radiation. A non-parametric wind regression method was used to highlight the sources of the air pollutants. The long-range transport of SO_2 and two hotspots for CO from traffic and from residential heating emissions were emphasized. The traffic hotspot situated north of the measuring site, close to the city ring road, is also a hotspot for NO_x. The air quality during the cold season was evaluated by comparing the measured gas concentration with the European limits. During the measuring period, the values for NO_2, CO, and SO_2 concentration were at least two times lower than the European Union pollution limits. Only twice during the study period was the concentration of O_3 higher than the established limits.

Keywords: atmospheric gases; sources; European regulations

---

1. Introduction

The atmospheric constituents, gases and aerosols have been intensely studied in the past decades [1–3] due to their dramatic influence the human life. The gases and aerosols are produced by natural and anthropological processes. According to the latest Intergovernmental Panel for Climate Change (IPCC) report [3] and United Nations Framework Convention on Climate (COP24) report [4], the predominant sources for greenhouse gases are 35% from energy industry for electricity and heat generation, 24% from agriculture, 21% from industry, 14% transport and 6% from buildings. The sources for aerosols, and especially for fine particles, are 25% from traffic, 15% from industries including the electricity generation sector, 20% from domestic fuel combustion, 22% from other human origin activities and only 18% from natural sources. Consequently, for both gases and aerosols, the anthropogenic sources account for approximately 70–80%.

Recent studies focused on the direct and indirect effect of gases and aerosols on radiative budget [1,3], human health [4], as well on biogeochemical cycles [3]. Atmospheric constituents can change the radiative balance of the Earth through cooling or heating the atmosphere. In recent years, not only the greenhouse gases (i.e., carbon dioxide (CO_2), methane (CH_4), ozone (O_3), nitrous oxide (N_2O), and fluorinated gases), which have a warming effect [5], but also the effect of minor gases are investigated [3]. It has been shown that volatile organic compounds (VOC) have a warming effect, sulfur dioxide (SO_2) has a cooling effect, while nitrogen oxides (NO_x) and aerosols can generate both effects [4]. In addition, it has been shown [2,6] that the indirect effect of atmospheric compounds (organic aerosol, nitrate, and sulfate) is influencing the cloud formation (i.e., atmospheric compound are acting as cloud condensation nuclei).

The impact of gases (carbon monoxide (CO), O_3, NO_x, abd VOC) and aerosols (e.g., brown carbon and organic aerosols including polyaromatic hydrocarbons) on health represent is a topic of high interest [4,7] because they can generate mainly respiratory and cardiovascular problems. The statistics [4] show that deaths related to air pollution are caused by ischemic heart disease (26%), strokes (24%), chronic obstructive pulmonary disease (43%) and lung cancer (29%). Recent studies are focused mainly on atmospheric evolution of gases composition based on global emission inventory and biogenic models [8]. The anthropogenic emissions, which are insufficiently characterized (e.g., those that are produced by residential heating) in some areas, are estimated to increase. Thus, they will have a high impact on human health if no regulations are adopted [2].

Due to this great impact, reducing atmospheric pollution represents a priority for society [3], requiring a deep understanding of the chemical and physical processes that lead to the formation or depletion of atmospheric compounds. The continuous interactions between gases and aerosols facilitated by meteorological parameters represent an important issue in reducing the concentration of air pollutants. For example, in the presence of OH radicals, VOC can oxidize nitric oxide (NO) to NO_x, thus increasing the O_3 production during the day. O_3 increases with the VOC concentration, while increasing the NO_x concentration may generate an increase or decrease of the O_3 level, depending on the existing ratio between VOC and NO_x [9].

The importance of the studies on atmospheric gases evolution was worldwide recognized in the last century [8,10]. In Europe, the majority of studies on atmospheric gas evolution focus on Northern, Southern, and Western Europe [10–12] and, to a lesser extent, on Eastern Europe [13–16]. The aim of this study was to provide, for the first time to the authors' knowledge, a comprehensive study of the gaseous atmospheric constituents in a peri-urban area in Romania, a country situated in Eastern Europe. In this study, gas reactions were evaluated during the cold season (1 December 2017–4 March 2018). The analysis for the three months period was focused on the temporal variation of NO_x, SO_2, O_3, CO, NO_2, NO, and CH_4, together with meteorological parameters.

This article is structured as follows. Section 2 describes the methodology, the data treatment, and processing. The diurnal behavior and the chemical reactions (as nighttime or daytime chemistry) dominant during winter together with the spatial source origin of the measured air pollutants are described in Section 3. Finally, Section 4 summarizes the results of the study.

2. Methodology

2.1. Measurement Site

The measurement site is located at the Romanian Atmospheric 3D Observatory (RADO) placed at Măgurele, a small residential city, situated 10 km south from Bucharest, the capital city of Romania, and approximately 2 km from the ring road of Bucharest. Măgurele is located in a peri-urban area influenced by urban sources and by agriculture activities from the neighboring areas [17–21]. The measurements were organized over a three months period from 1 December 2017 to 4 March 2018 as a part of the EMEP (European Monitoring and Evaluation Programme)/ACTRIS (Aerosol, Clouds and Trace Gases Research Infrastructure)/COLOSSAL(Chemical On-Line cOmpoSition and Source Apportionment of fine aerosoL-Cost action) international campaign that involved 57 sites in 24 countries all over Europe. The aim of this wintertime intensive measurement campaign was to evaluate the gases concentration used as tracers for the emission sources.

2.2. Measurements

The gases sampled during campaign were: NO, NO_2, NO_x, O_3, SO_2, CH_4, and CO. In addition, the meteorological parameters were also measured (i.e., temperature, wind speed, wind direction, relative humidity, solar radiation, and rainfall). For gases sampling, Horiba monitors (370 Horiba) with standard measurement principles (SR EN 14212, SR EN 14211, SR EN 14625, SR EN 14626, and SR EN 12619) were used. The CO monitor uses a non-dispersive infrared absorption with cross flow modulation technology. The SO_2 is measured using UV fluorescence: a Xe lamp is used to excite the molecules which emit a characteristic fluorescent signal in the range 220–420 nm. The NO_x monitor is based on a cross flow modulation chemiluminescence method, using the reaction between NO and O_3 to form excited molecules of NO_2^*. Returning to the ground state, the chemoluminscence signal, emitted is in the range 600–3000 nm, is proportional to the concentration of the NO. For NO_2 measurements, a converter is used to transform NO_2 to NO. The CH_4 monitor uses a flame ionization detection method with selective combustion.

The Horiba analyzers were calibrated before the start of the campaign using the recommended procedures by the manufacturer. Every measured point was selected for the analysis after passing the validity and error status flags implemented in the data acquisition software. Outliers (defined as points that were measured for a period less than ten minutes with an increased concentration by a factor of 3 relative to the ten-minute average of the measured neighbor points) were manually removed. The gas data were sampled every 3 min and were averaged to the same time stamp as the meteorological parameters at each 10 min or at different time intervals (i.e., 1, 8 or 24 h) to compare with the European limits. The gases data matrix was re-gridded using the time from the meteorological measurements as data stamp. For every three gases samples, 1 min data were obtained through linear interpolation and then averaged at 10 min to obtain the same data stamp as meteorological parameters.

Meteorological data were sampled with a 10 min resolution, at 15 m above the ground, using the meteorological sensors (temperature, wind direction, relative humidity, solar radiation, and rainfall) integrated in the Environmental data Compact Weather Data Recording System. The wind speed data were provided by the National Environmental Agency, from the sampling site in Măgurele, situated 100 m from the RADO site.

2.3. Data Treatment and Processing

The diurnal trend and weekday variation were computed for NO, NO_2, NO_x, O_3, SO_2, CH_4, and CO to highlight the specific variations and the local sources. The non-parametric wind regression (NWR) method was used to evaluate the spatial distribution of the pollutants' concentration [22] and to assess the local and long range transported sources. This technique uses wind data and concentration of measured chemical species to identify wind speed and direction that are associated with high concentrations, thus indicating the presence of a source. As described by Henry et al. [23], for every angle θ (wind direction) and v (wind speed), an expected concentration is calculated by weighting the measured concentration with the following Kernel functions

$$K_1(x) = \frac{1}{\sqrt{2\pi}} e^{-0.5 \cdot x^2} \qquad -\infty < x < \infty \tag{1}$$

$$K_2(x) = \begin{cases} 0.75\left(1 - x^2\right), -1 < x < 1 \\ 0, \text{otherwise.} \end{cases} \tag{2}$$

The expected concentration for every pair, $E\left(C|\theta, v\right)$ is calculated using the following relation

$$E\left(C|\theta, v\right) = \frac{\sum_{i=1}^{N} K_1\left(\frac{\theta - w_k}{\sigma}\right) K_2\left(\frac{v - s_j}{u}\right) C_i}{\sum_{i=1}^{N} K_1\left(\frac{\theta - w_i}{\sigma}\right) K_2\left(\frac{v - s_i}{u}\right)} \tag{3}$$

where w_i and s_i are the measured wind direction and speed, respectively; C_i is the concentration of the pollutant; N is the total number of data points; and σ and u are smoothing parameters. Then, the expected concentration for every pair is weighted by the frequency of the wind parameters and the final mean concentration, $S(\theta, U)$, associated with the wind speed in the interval $V = [v_1, v_2]$ and the wind direction in the interval $W = [\theta_1, \theta_2]$, is

$$S(\theta, U) = \frac{\Delta v \Delta \theta}{N \sigma u} \sum_{\theta_k = \theta_1}^{\theta_2} \sum_{v_j = v_1}^{v_2} \sum_{i=1}^{N} K_1\left(\frac{\theta - w_i}{\sigma}\right) K_2\left(\frac{v - s_i}{u}\right) C_i \tag{4}$$

For the NWR method, ZeFir [22] was used. The software is freely available at https://sites.google.com/site/zefirproject/home (accessed on 1 May 2019). In our study, the smoothing parameters used were: angle smooth of 22.98 and radial smooth of 3, as automatically estimated by the Zefir software.

3. Results

3.1. The Diurnal Behavior and Chemical Reactions

The campaign period was characterized by an average temperature of 2.7 °C, typical for a warm winter, with the temperature ranging from -15 to 17.5 °C. The negative values for temperature represented only 22.5% of all measured values. The diurnal pattern of the temperature highlights the minimum value (0.33 °C) at 05:00 UTC, while the highest value (15.7 °C) was recorded around 13:00 UTC. The relative humidity during the measurements period was high, the average for the entire period being 91%, while the diurnal trend varying 78–98%. During the entire campaign, the wind speed had an average value of 1 m s^{-1}, which is characteristic for calm conditions, and varied between 0 and 4.9 m s^{-1}.

The diurnal trend of solar radiation showed a maximum peak of 44 W m^{-2} at 11:00 UTC having an average for the entire period of 10 W m^{-2}. The solar radiation represents an important factor that modulates the O$_3$ formation. The time series of the solar radiation was well correlated with the O$_3$ concentration (Figure 1). The highest concentration for O$_3$ (i.e., 76 ppb) was reached 2 h after the maximum value of the solar radiation of 126 W m^{-2} on 1 March 2018 at 11:20 UTC. In Figure 1, the anti-correlation between NO$_x$ and O$_3$ can also be observed, the direct reactions between these two gases being responsible for this behavior.

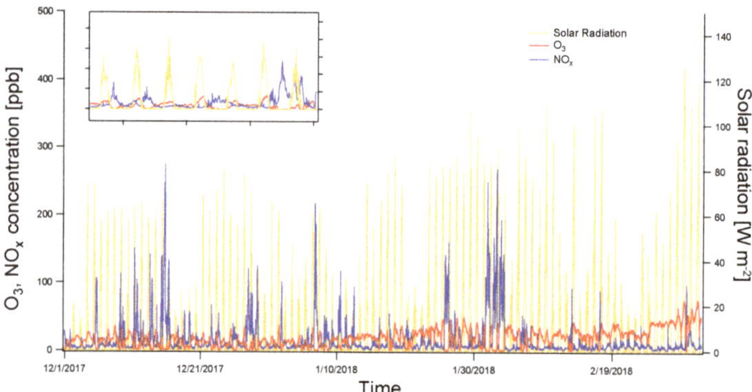

Figure 1. Time series of hourly average concentration for O$_3$ (red), NO$_x$ (blue) and solar radiation (yellow) during winter the period (1 December 2017–4 March 2018). The inset in the left corner emphasizes the correlations between solar radiation and O$_3$ and anti-correlation among solar radiation, O$_3$ and NO$_x$.

The reactions in troposphere that describe the relations between NO_x and O_3 are [11,24,25]

$$NO_2 + h\nu \ (\lambda < 0.424 \ \mu m) \rightarrow NO + O^* \tag{5}$$

$$O^* + O_2 \rightarrow O_3 + M \tag{6}$$

$$O_3 + NO \rightarrow NO_2 + O_2 \tag{7}$$

where M is a molecule (N_2 or O_2) that absorbs the vibrational energy in excess. The reactions represent a null cycle, the total mixing ratio of NO_x and OX ($NO_2 + O_3$) remain unchanged, but allows the partitioning between their components (NO and NO_2 for NO_x, and NO_2 and O_3 for OX). During daytime, NO_2, NO, and O_3 are in equilibrium for a few moments forming a photo-stationary state. For the stationary state, the components are linked through the following relation

$$NO \times O_3 = \frac{J_2}{k_1} \times NO_2 \tag{8}$$

where J_2 is the rate for NO_2 photolysis and k_1 is the rate coefficient for the reaction of NO with O_3 and it depends exponentially on the temperature [26]

$$k_1 \left(ppm^{-1} \ min^{-1} \right) = 3.23 \times 10^3 \exp\left(-\frac{1430}{T} \right) \tag{9}$$

These chemical pathways lead to the in situ O_3 production in the troposphere through the NO_x reactions, thus O_3 can be considered secondary pollutant. NO and NO_2 concentrations had a peak in the morning, corresponding to the morning traffic hour and when a minimum in O_3 concentration was observed (Figure 2). During daytime, NO decreased with the increase of solar radiation, reaching a minimum value around 13:00 UTC. The oxidation of NO to form NO_2, according to (7), was also observed in the measured data, the NO_2 concentrations being higher than the NO concentrations all day, except the morning hour peak, when the emission of NO occurred. The NO_x emissions from combustion processes were formed mainly (90%) from nitric oxide and the rest was nitrogen dioxide [27]. Thus, NO was the primary pollutant and NO_2 and O_3 were secondary products. The Pearson correlation coefficients between NO_x and NO_2 was 0.8, while for NO_x and NO was 0.98.

Using the NWR model, the NO source was identified as situated north of the sampling site, where the Bucharest ring road represents the main contributor (Figure 2). NO_2 had a similar pattern to NO, the concentration being more widely distributed in space, due to a longer residence time compared with NO, and it was transported longer distances before producing O_3. The NO_x source plot (Figure S2) emphasizes the sources for NO and NO_2 in the northern part.

The maximum value in solar radiation corresponded to a minimum in NO_2 concentration, while the maximum O_3 concentration was reached 1 h after the peak in solar radiation. Starting from 06:00 UTC, the photochemical reactions were enhanced, since the solar radiation started to increase. A negative correlation between O_3 and NO_x was observed, the Pearson coefficient being -0.44, as ozone diminished while NO_x increased [9].

The weekly diurnal (Figure S3) of NO, NO_2 and NO_x showed an increasing trend during the week, the maximum concentration being reach on Friday, while the minimum concentration for NO_x being on Sunday, corresponding with the maximum concentration for O_3. The authors speculate that Friday represents the start of the weekend with usually more intense traffic on the Bucharest ring road and Saturday is also characterized by high traffic, more related with recreational activities (e.g., shopping), unlike Sunday when traffic is reduced and the activities are mainly indoors.

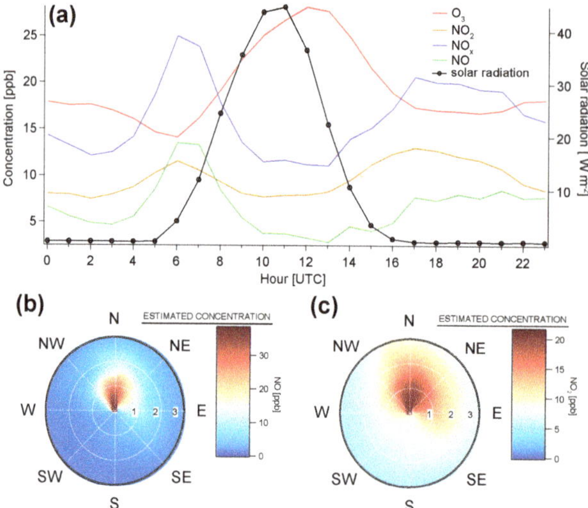

Figure 2. (**a**) Diurnal trend for ozone (red line), nitrogen dioxide (orange line), nitric oxide (green line), nitrogen oxides (blue line) and solar radiation (black line), computed for the cold season between 1 December 2017 and 4 March 2018. Source estimations for: (**b**) NO (ppb, shaded according to the scale); and (**c**) NO_2 (ppb, shaded according to the scale). For the source estimation plots, the inner white circles represent the wind speed in m s^{-1}.

Ozone presented an opposite weekly variation compared with NO_x. The minimum O_3 concentration (17 ppb) was reached on Friday, corresponding to the maximum concentration for NO_x (25 ppb). The maximum concentration for O_3 (23 ppb) was on Sunday, corresponding to a minimum NO_x (7 ppb). This behavior explains the correlation and chemical reaction between NO_x and O_3, which was also underlined by the diurnal pattern.

CH_4 and CO sources are diverse, being associated with combustion (fossil fuel or biomass burning) or with microbiological activity (e.g., from waste deposition for methane) [28]. In our case, one of the main sources for CH_4 and CO was represented by exhaust engines. The diurnal patterns of CH_4 and CO showed similar trends as the NO_x diurnal variation, with higher concentration during the morning (05:00–06:00 UTC) and evening (around 19:00 UTC) (Figure 3). The diurnal pattern of CO presented a peak at 06:00 UTC of 0.42 ppm, but this value was lower than the midnight concentration, which was around 0.45 ppm. After 15:00 UTC, the CO concentration started to increase up to 0.51 ppm, the maximum evening peak. This behavior for CO can be explained by the diurnal planetary boundary layer (PBL) evolution and the presence of two main sources, traffic and residential heating. The peak in the morning was more influenced by the traffic, while, in the evening, besides the traffic, residential heating contributed as well. This explains why the peak concentration in the second part of the day appeared at a later hour. These sources for CO were highlighted also by the NWR model, the estimated concentration for CO being higher in the northern part, but other minor sources contributed as well in the surrounding zone of the measuring site (Figure 3), probably produced by residential heating. This hypothesis is reinforced by the large number of houses in that area. According to the 2011 Annual Census (Ilfov Annual Census, 2011, http://www.prefecturailfov.ro/Biroul-de-Presa/Comunicat%20-%20DATE%20PROVIZORII% 20RPL%202011_JUDET_Ilfov.pdf, accessed on 5 August 2019), in Măgurele, 3376 housing units are distributed over 45 km^2.

For CH_4, the diurnal analysis emphasized higher concentration in the first part of the day, with a morning peak of 2.72 ppm at 05:00 UTC. An evening peak could also be identified, the concentration being around 2.62 ppm at 19:00–20:00 UTC. This behavior with higher concentrations in the morning and in the evening were similar to the NO_x diurnal pattern, indicating a traffic source for CH_4 as well [29]. The source for CH_4 in our case, derived from the NWR model (Figure 3), emphasized higher concentrations in N and the ESE part, possibly due to traffic to the N and a landfill situated 7 km from the sampling site to the SE. This landfill is one of the largest landfills in the southeast of the country with a total surface of 38.6 ha. The capacity of this landfill is 486,384.68 t year^{-1}, with a daily flow of 2400 t of waste. The frequency of waste disposal is one garbage truck every 5 min. This information was obtained from an online resource (https://ecosud.ro/en/depozitul-ecologic-vidra-bucuresti/, accessed on 31 July 2019). The emission of CH_4 from the solid waste disposal has been previously described [28,30].

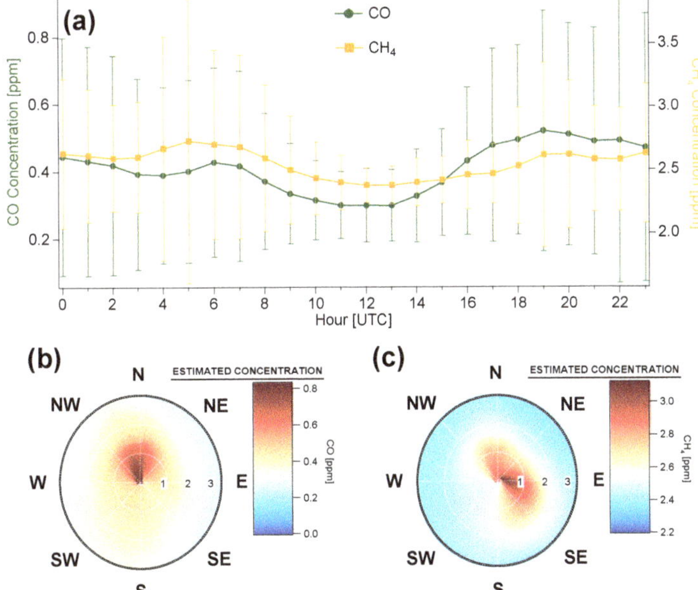

Figure 3. (**a**) Diurnal pattern computed for cold season between 1 December 2017 and 4 March 2018 of carbon monoxide (green line) and methane (yellow line) emphasizing peaks during morning and evening hours. The vertical error bars represent the standard deviation from the mean value. Source estimations are shown for: (**b**) CO; and (**c**) CH_4. For the source estimation plots, the inner white circles represent the wind speed in m s^{-1}.

The minimum concentration for CO and CH_4 occurred at noon, corresponding to maximum values for solar radiation and PBL height. These patterns can be explained by the OH oxidation described by the following equations [24].

$$OH + CO \rightarrow CO_2 + H \tag{10}$$

$$OH + CH_4 \rightarrow CH_3 + H_2O \tag{11}$$

In the atmosphere, OH is produced from the photooxidation of the ozone and followed by the reaction with water. The OH diurnal trend, with a maximum value at noon [31,32] and the diurnal trend of CO and CH_4 support the OH oxidation and the formation of CO_2 and CH_3. The pathway for methane oxidation continues and produces water vapor and CO that is oxidized to CO_2, thus the oxidation of CH_4 generates other greenhouse gases (CO_2, stratospheric water vapor, and O_3) and important atmospheric oxidants (O_3 and OH) [33]. In addition, CH_4 is the second most important greenhouse gas after CO_2, and its lifetime is 8–10 years [33]. Most CH_4 is removed through OH oxidation and only a small fraction is removed through surface deposition [33]. Thus, CH_4 oxidation represents a method for OH loss and any increase in the CH_4 concentrations will suppress OH and increase the CH_4 lifetime and concentration [33].

CO and CH_4 can also contribute to O_3 formation through photochemical processes that could be more important during summer. The correlations between O_3 and CO could explain the production or consumption of O_3 [34] depending on the intensity of photochemical processes. The negative correlation suggests O_3 destruction through photochemical reactions or CO production. The moderate correlations and small slopes showed the incomplete photooxidation of fresh emitted plumes, which did not reach the O_3 production potential. In our study, the Pearson coefficient for O_3 and CO had a negative value of -0.47, similar to other winter measurements [35], possibly due to the O_3 destruction through the NO_x titration. CO did not directly affect the O_3 concentration but was emitted together with NO_x (the Pearson coefficient of CO and NO_x was 0.73). Thus, the negative correlation between O_3 and CO could be explained by the NO_x-O_3 chemistry [36]. The minimum concentration for CO, 0.3 ppm during 11:00–13:00 UTC, coincided with the maximum value for ozone, approximately 28 ppb.

Weekly variation of CO concentrations (Figure S4) did not show a specific pattern, except for a slight increase on Friday, with a mean concentration of 0.48 ppm and a minimum concentration of 0.36 ppm on Sunday. This trend, with higher concentrations on Friday, which had the same appearance as NO_x as well as the Pearson coefficient between CO and NO_x of 0.73, suggest that they can be produced by same source.

Another tracer used in our study was SO_2, which can be emitted by biomass burning as a minor source [37], volcanoes emissions and anthropogenic activities such as fossil fuel combustion [24]. In addition, it can be formed by the oxidation of other sulfur gases such as H_2S, DMS, COS, and CS_2 produced by soils, plants, marshlands and oceans due to phytoplankton activity [24]. The reactions in the atmosphere, which involve the oxidation of SO_2 mainly by the OH radical and only in aqueous phase, are as follow [24]:

$$SO_2 + OH + M \rightarrow HSO_3 + M \tag{12}$$

$$HSO_3 + O_3 \rightarrow HO_2 + SO_3 \tag{13}$$

$$SO_3 + H_2O \rightarrow H_2SO_4 \tag{14}$$

The measured SO_2 diurnal variation during the winter campaign had a peak at noon, the maximum concentration being recorded at 12:00 UTC with a value of 2.6 ppb (Figure 4). The diurnal pattern of SO_2 in Figure 4 cannot be explained by the OH oxidation reactions, but could be explained by the multi-step reaction of H_2S with the OH radical [24]. In the atmosphere, H_2S is produced by marshlands and reactions in soil [24].

$$OH \rightarrow HS, H_2O \rightarrow SO_2 \tag{15}$$

In our case, the location of the sampling site, surrounded by agricultural areas, could be insufficient to explain this pattern due the lack of agriculture activities during the cold season. Several other cases with noontime peaks of the SO_2 diurnal pattern have been observed around the world. Adame et al. [11] observed occasionally maximum values of SO_2 concentration during noontime in central Spain attributed to transport from industrial areas. In the North China Plain, the case of noontime SO_2 peak is more often, with an occurrence of 50–75% at four different sites during more than a year measurements data [38]. The authors attributed this pattern to four possible cases: down-mixing of high concentrations in the planetary boundary layer, influence from the mountain–valley breezes, severe fog and haze events, and transport of plumes. In our case, the influence of mountain or valley winds does not apply as the measuring site is situated on a plain (i.e., Bărăgan Plain) and the mountains are approximately 100 km away from the measuring site. The hypothesis of mixing in the PBL and the influence of severe fog or haze events would have caused different behavior in other gases measured during the intensive measurement campaign as well. The only possible cause could be transport from different regions. This hypothesis was validated by the source identification, the model showing a hotspot in the W–SW associated with wind speeds higher that 2 m s^{-1}, suggesting the transport of SO_2 from other regions (Figure 4).

The weekly variation of SO_2 (Figure S4) showed the maximum value (2.2 ppb) on Friday, similar to NO_x and CO, while the minimum concentration was about 1.5 ppb on Wednesday.

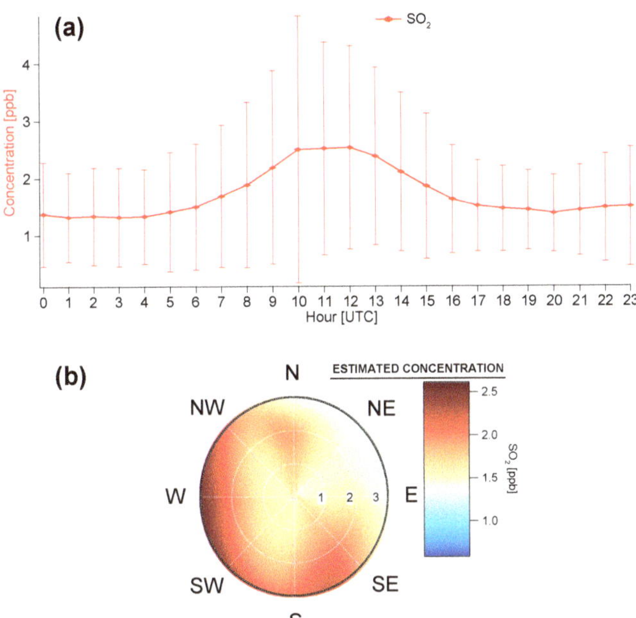

Figure 4. (**a**) Diurnal pattern computed for cold season between 1 December 2017 and 4 March 2018 for sulfur dioxide. The vertical error bars represent the standard deviation from the mean value. (**b**) Long-range transport was highlighted by NWR model for source estimation; higher concentrations were correlated with higher wind speeds. For the source estimation plots, the inner white circles represent the wind speed in m s^{-1}.

3.2. Assessment of Time Series Concentrations Relative to the European Regulations

The average, maximum and minimum values for all parameters measured during the cold season campaign are shown in Table 1. The O_3 mean concentration had a value of 19.45 ppb, lower that the annual mean concentration reported in the United Kingdom in 2017 (25.7 ppb) [39]. The highest concentration for O_3 reported in Bucharest was 0.958 ppm in an urban area with intense traffic, while in the rural area the maximum value measured was 0.250 ppm [13].

The SO_2 mean value for the winter campaign was 1.71 ppb and the maximum was 23 ppb. The mean value was lower than the previous values reported in a traffic area in Bucharest, 12 ppb [13], or in other cities in Europe: 3 ppb in Spain [11], and around 2 ppb, as the lowest value report during a two-week winter campaign in Prague [10]. For nitrogen oxides, the mean value was 16.3 ppb, of which the higher proportion was NO_2 (60%) with an average of 9.7 ppb. However, the maximum value recorded was higher for NO, 237 ppb, because it had a greater proportion during primary emission. In the literature, the NO_x values reported are 70 ppb for Bucharest traffic area [13], and from 34 to 52 ppb monthly average during winter in China [40]. The CO measured in the present study had an average concentration of 0.4 ppm, while the maximum value during the campaign was 11 ppm, higher than the maximum measured previously in Bucharest (1.54 ppm) [13].

Table 1. Average, minimum and maximum concentrations of the parameters derived.

Parameter Derived-Unit	Average Concentration for the Entire Period	Minimum Concentration for the Entire Period	Maximum Concentration for the Entire Period
O_3 (ppb)	19.45	0.5 (detection limit)	76
SO_2 (ppb)	1.71	0.5 (detection limit)	23
NO (ppb)	6.6	0.5 (detection limit)	237
NO_2 (ppb)	9.7	1.31	60
NO_x (ppb)	16.3	1	274
CO (ppm)	0.4	0.1	11
CH_4 (ppm)	2.55	2.1	13

The great impact that air pollutants have on human heath require implementation of a legislation framework that which establishes limits for accepted air pollution concentration. These standards are adopted on global and/or regional scale and include temporal maximum concentration values for different pollutants. At European level, the limits are established through European Directives, being implemented in national legislation. In this study, an important task was the comparison of the measured values with the existent European limits, transposed in national legislation, for air pollutants. The limits are set for an averaging period of 1, 8 or 24 h and in some cases this limit can be exceeded for only a limited number of times per year (Table 2). For these three-month measurements, only ozone exceeded the limit (Figure 5), having two values higher than 60 ppb on 1 March and 3 March 2018 and one value close to the limit, 59.3 ppb, also on 3 March 2018. The exceeding of the allowed ozone concentration limit was correlated with solar radiation increases. On these two days, solar radiation reached the maximum values of 126 and 119 W m^{-2}, respectively, recorded over entire period of measurements. The maximum NO_2 average value, 58 ppb, was almost two times lower than the established limit, while the maximum CO average was almost five times lower and the maximum SO_2 value was ten times lower than the specified limits in the legislation. In conclusion, the average concentrations of these parameters were in accordance with the site location. According to the National Environmental Agency, the sampling site in Măgurele is defined as a rural location (http://www.calitateaer.ro/, accessed on 9 July 2019), with expected lower concentrations.

Table 2. European standards for air pollutants.

Pollutant	Limit Concentration	Averaging Period	Accepted Exceedances per Year
O_3	120 µg m^{-3} (60 ppb)	maximum daily 8 h mean	25 days in 3 years
SO_2	125 µg m^{-3} (47 ppb) 350 µg m^{-3} (134 ppb)	24 h 1 h	3 24
CO	10 mg m^{-3} (8 ppm)	maximum daily 8 h mean	N/A
NO_2	200 µg m^{-3} (105 ppb) 40 µg m^{-3} (21 ppb)	1 h 1 year	18 N/A

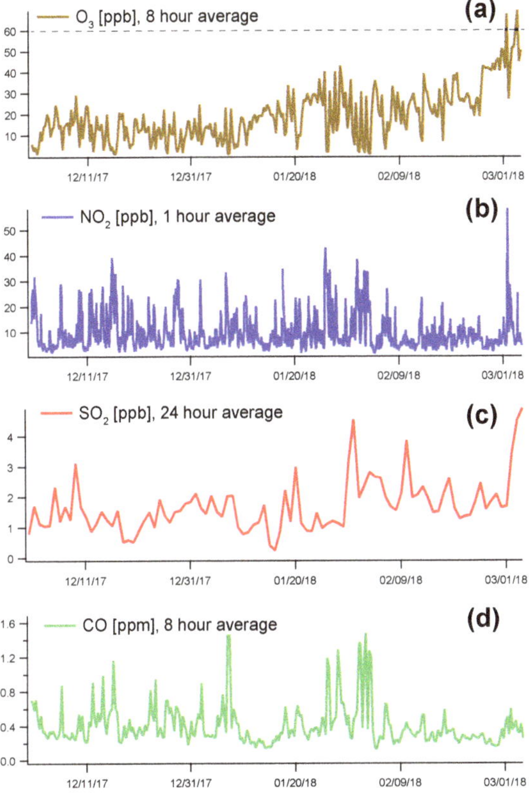

Figure 5. Time series air pollutants averaged according to the averaging period specified in the European Union legislation (indicated on each panel) for: (**a**) O_3; (**b**) NO_2; (**c**) SO_2; and (**d**) CO concentration. The horizontal line in (**a**) represents the limit value for O_3 according to the European Union legislation.

4. Conclusions

The winter campaign was deployed over three months between December 2017 and March 2018, being characterized by mainly positive temperatures, representing 77.5% of the total recorded values. Solar radiation represented the driving force in gases chemistry, especially for O_3 and NO_x formation and depletion.

Different behaviors during weekend and weekdays were observed for gases produced by exhaust engines. The highest concentration for NO_x was recorded on Friday, while the maximum concentration for O_3 was recorded on Sunday, explained by the chemical reactions between NO_x and O_3.

The average concentrations of the measured gases were at least two times lower than legal limits established by European commission: NO_2 maximum of 58 ppb was almost two times lower than the required limit; the maximum CO average was almost five times lower; and the maximum SO_2 value was ten times lower than the specified limits in the legislation. Only ozone concentration exceeded the 60 ppb EU limit on two different days (i.e., 1 March 2018 and 3 March 2018).

The influence of different sources was evidenced using the NWR model. The main contributor for our sampling site is the Bucharest ring road, as clear evidenced in the NO_x model distribution and CO concentrations. Other sources found were related to residential heating, highlighted by CO estimated concentration distribution. The CO model indicated the influence mainly of the northern part, where the ring road of Bucharest is located, and the influence of surrounding living areas was evidenced as well. The presence of a landfill that services Bucharest city located in the E–SE part was highlighted by CH_4 estimated concentration. Another source was related to the long-range transport: the SO_2 concentration increased mainly with the wind speed.

This work presents the gases constituents behavior during the cold season at the measuring site in Măgurele. Future work is required to describe the diurnal patterns and reactions of gases during warm season, as photooxidation processes are more important during summer. In addition, the source estimation during other seasons would represent a great interest as the measuring site could be influenced by different sources resulted from agriculture activities (during spring and autumn) or by long range transport pollution from wild fires (during summer).

An important objective of the present study was the evaluation of the air quality compared with the European limits for the measured pollutants. The lacking information about the air quality in Romania could be covered by a future study for evaluating the pollutants' concentrations in different environments (i.e., urban, regional, and background). In addition, multi-annual statistics would show the effect of national legislation on emissions rates and if further actions are required in order to fulfill the European Union recommendations.

Supplementary Materials: The following are available online at http://www.mdpi.com/2073-4433/10/8/478/s1, Figure S1: Diurnal trend for ozone, nitrogen dioxide, nitric oxide, nitrogen oxides and solar radiation; Figure S2: Source estimation for NO_x; Figure S3: Weekly variation for ozone, nitrogen dioxide, nitric oxide, and nitrogen oxides; Figure S4: Weekly variation for sulfur dioxide, carbon monoxide and methane.

Author Contributions: Conceptualization, C.A.M. and L.M.; Methodology, C.A.M., L.M., C.R., A.D. and F.T. Formal Analysis, C.A.M. and L.M.; Investigation, all authors; Writing—Original Draft Preparation, C.A.M.; Writing—Review and Editing, all authors; Visualization, C.A.M. and B.A.; and Supervision, L.M. and C.S.

Funding: This research was funded by the Ministry of Research and Innovation through Program I—Development of the national research-development system, Subprogram 1.2—Institutional Performance—Projects of Excellence Financing in RDI, Contract No.19PFE/17.10.2018; by the Romanian National Core Program Contract No.18N/2019; and by ROSA-STAR Ctr. 162/20.07.2017, Ctr. 136/20.07.2017.

Acknowledgments: The authors gratefully acknowledge the Zefir software and the Romanian National Environmental Agency for providing data. We thank Jeni Vasilescu and Dragoş Ene from the National Institute of Research and Development for Optoelectronics INOE 2000 for their comments on an early version of the manuscript.

Conflicts of Interest: The authors declare no conflict of interest.

References

1. Solomon, S.; Qin, D.; Manning, M.; Chen, Z.; Marquis, M.; Averyt, K.B.; Tignor, M.; Miller, H.L. (Eds.) *Climate Change 2007: The Physical Science Basis. Contribution of Working Group I to the Fourth Assessment Report of the Intergovernmental Panel on Climate Change;* Cambridge University Press: Cambridge, UK; New York, NY, USA, 2007.

2. Stocker, T.F.; Qin, D.; Plattner, G.-K.; Tignor, M.; Allen, S.K.; Boschung, J.; Nauels, A.; Xia, Y.; Bex, V.; Midgley, P.M. (Eds.) *Climate Change 2013: The Physical Science Basis. Contribution of Working Group I to the Fifth Assessment Report of the Intergovernmental Panel on Climate Change;* Cambridge University Press: Cambridge, UK; New York, NY, USA, 2013; p. 1535.

3. IPCC. *Global Warming of 1.5 °C. An IPCC Special Report on the Impacts of Global Warming of 1.5 °C Above Pre-Industrial Levels and Related Global Greenhouse Gas Emission Pathways, in the Context of Strengthening the Global Response to the Threat of Climate Change, Sustainable Development, and Efforts to Eradicate Poverty;* Masson-Delmotte, V., Zhai, P., Pörtner, H.O., Roberts, D., Skea, J., Shukla, P.R., Pirani, A., Moufouma-Okia, W., Péan, C., Pidcock, R., et al., Eds.; IPCC: Geneva, Switzerland, 2018, in Press.

4. COP24 Special Report. Health and Climate Change. In Proceedings of the 24th Conference of the Parties to the United Nations Framework Convention on Climate Change (COP24), Katowice, Poland, 2–15 December 2018; 38p.

5. Montzka, S.A.; Dlugokencky, E.J.; Butler, J.H. Non-CO_2 greenhouse gases and climate change. *Nature* **2011**, *476*, 43–50. [CrossRef] [PubMed]

6. Boucher, O.; Randall, D.; Artaxo, P.; Bretherton, C.; Feingold, G.; Forster, P.; Kerminen, V.-M.; Kondo, Y.; Liao, H.; Lohmann, U.; et al. Clouds and Aerosols. In *Climate Change 2013: The Physical Science Basis. Contribution of Working Group I to the Fifth Assessment Report of the Intergovernmental Panel on Climate Change*; Stocker, T.F., Qin, D., Plattner, G.-K., Tignor, M., Allen, S.K., Boschung, J., Nauels, A., Xia, Y., Bex, V., Midgley, P.M., Eds.; Cambridge University Press: Cambridge, UK; New York, NY, USA, 2013.

7. Crippa, M.; Janssens-Maenhout, G.; Guizzardi, D.; Van Dingenen, R.; Dentener, F. Contribution and uncertainty of sectorial and regional emissions to regional and global $PM_{2.5}$ health impacts. *Atmos. Chem. Phys.* **2019**, *19*, 5165–5186. [CrossRef]

8. Huszar, P.; Belda, M.; Halenka, T. On the long-term impact of emissions from central European cities on regional air quality. *Atmos. Chem. Phys.* **2016**, *16*, 1331–1352. [CrossRef]

9. Pancholi, P.; Kumar, A.; Bikundia, D.S.; Chourasiya, S. An observation of seasonal and diurnal behavior of O_3–NO_x relationships and local/regional oxidant (OX = O_3 + NO_2) levels at a semi-arid urban site of western India. *Sustain. Environ. Res.* **2018**, *28*, 79–89. [CrossRef]

10. Thimmaiah, D.; Hovorka, J.; Hopke, P.K. Source apportionment of winter submicron Prague aerosols from combined particle number size distribution and gaseous composition data. *Aerosol. Air Qual. Res.* **2009**, *9*, 209–236. [CrossRef]

11. Adame, J.; Notario, A.; Villanueva, F.; Albaladejo, J. Application of cluster analysis to surface ozone, NO_2 and SO_2 daily patterns in an industrial area in Central-Southern Spain measured with a DOAS system. *Sci. Total. Environ.* **2012**, *429*, 281–291. [CrossRef]

12. Clapp, L.J.; Jenkin, M.E. Analysis of the relationship between ambient levels of O_3, NO_2 and NO as a function of NO_X in the UK. *Atmos. Environ.* **2001**, *35*, 6391–6405. [CrossRef]

13. Apascariţei, M.; Popescu, F.; Ionel, I. Air pollution level in urban region of Bucharest and in rural region. In Proceedings of the 11th WSEAS International Conference on Sustainability in Science Engineering, Timişoara, Romania, 27–29 May 2009.

14. Ionel, I.; Nicolae, D.; Popescu, F.; Talianu, C.; Belegante, L.; Apostol, G. Measuring air pollutants in an international Romanian airport with point and open path instruments. *Rom. J. Phys.* **2011**, *56*, 507–519.

15. Popescu, F.; Ionel, I.; Lontis, N.; Calin, L.; Dungan, I.L. Air quality monitoring in an urban agglomeration. *Rom. J. Phys.* **2011**, *56*, 495–506.

16. Antonescu, B.; Ştefan, S. The urban effect on the cloud-to-ground lightning activity in the Bucharest area. *Rom. Rep. Phys.* **2011**, *63*, 535–542.

17. Mărmureanu, L.; Vasilescu, J.; Marin, C.; Ene, D. Aerosol source assessment based on organic chemical markers. *Rev.Chim.* **2017**, *68*, 853–857.

18. Dandocsi, A.; Nemuc, A.; Marin, C.; Andrei, S. Measurements of aerosols and trace gases in southern Romania. *Rev. Chim.* **2017**, *68*, 873–878.

19. Vasilescu, J.; Mărmureanu, L.; Nemuc, A.; Nicolae, D.; Talianu, C. Seasonal variation of the aerosol chemical composition in a Romanian peri-urban area. *Environ. Eng. Manag. J.* **2017**, *16*, 2491–2496.

20. Mărmureanu, L.; Vasilescu, J.; Ştefănie, H.; Talianu, C. Chemical and optical characterization of submicronic aerosol sources. *Environ. Eng. Manag. J.* **2017**, *16*, 2165–2172. [CrossRef]

21. Marin, C.; Stan, C.; Preda, L.; Marmureanu, L.; Belegante, L.; Cristescu, C.P. Multifractal cross correlation analysis between aerosols and meteorological data. *Rom. J. Phys.* **2018**, *63*, 805.

22. Petit; J.-E.; Favez, O.; Albinet, A.; Canonaco, F. A user-friendly tool for comprehensive evaluation of the geographical origins of atmospheric pollution: Wind and trajectory analyses. *Environ. Model. Softw.* **2017**, *88C*, 183–187.

23. Henry, R.; Norris, G.A.; Vedantham, R.; Turner, J.R. Source region identification using kernel smoothing. *Environ. Sci. Technol.* **2009**, *43*, 4090–4097. [CrossRef] [PubMed]

24. Hobbs, P. *Introduction to Atmospheric Chemistry*; Cambridge University Press: Cambridge, UK, 2000; p. 262.

25. Seinfeld, J.; Pandis, S. *Atmospheric Chemistry and Physics: From Air Pollution to Climate Change*, 2nd ed.; Wiley-Interscience: Hoboken, NJ, USA, 2006; p. 1152.

26. Mazzeo, N.A.; Venegas, L.E.; Choren, H. Analysis of NO, NO_2, O_3 and NO_x concentrations measured at a green area of Buenos Aires City during wintertime. *Atmos. Environ.* **2005**, *39*, 3055–3068. [CrossRef]

27. WHO guidelines for indoor air quality: Selected pollutants. Geneva 2010 World Health Organization. Chapter 5–Nitrogen dioxide. In *WHO Guidelines for Indoor Air Quality: Selected Pollutants*; World Health Organization, Regional Office for Europe: Geneva, Switzerland, 2010; p. 484.

28. Lee, U.; Han, J.; Wang, M. Evaluation of landfill gas emissions from municipal solid waste landfills for the life-cycle. *J. Clean. Prod.* **2017**, *166*, 335–342. [CrossRef]

29. Lipman, T.; Delucchi, M.A. Emissions of nitrous oxide and methane from conventional and alternative fuel motor vehicle. *Clim. Chang.* **2002**, *53*, 477–516. [CrossRef]

30. Bogner, J.; Spokas, K. Landfill CH4: Rates, fates, and role in global carbon cycle. *Chemosphere* **1993**, *26*, 369–386. [CrossRef]

31. Zanis, P.; Monks, P.S.; Schuepbach, E.; Penkett, S.A. The role of in situ photochemistry in the control of ozone during Spring at the Jungfraujoch (3580 m asl)–Comparison of model results with measurements. *J. Atmos. Chem.* **2000**, *37*, 1–27. [CrossRef]

32. Stone, D.; Whalley, L.K.; Heard, D.E. Tropospheric OH and HO$_2$ radicals: Field measurements and model comparisons. *Chem. Soc. Rev.* **2012**, *41*, 6348–6404. [CrossRef] [PubMed]

33. Isaksen, I.S.A.; Gauss, M.; Myhre, K.; Anthony, K.M.W.; Ruppel, C. Strong atmospheric chemistry feedback to climate warming from Arctic methane emissions. *Glob. Biogeochem. Cycles.* **2011**, *25*, GB2002. [CrossRef]

34. Choi, H.D.; Liu, H.; Crawford, J.H.; Considine, D.B.; Allen, D.J.; Duncan, B.N.; Horowitz, L.W.; Rodriguez, J.M.; Strahan, S.E.; Zhang, L.; et al. Global O$_3$—CO correlations in a chemistry and transport model. *Atmos. Chem. Phys.* **2017**, *17*, 8429–8452. [CrossRef]

35. Voulgarakis, A.; Telford, P.J.; Aghedo, A.M.; Braesicke, P.; Faluvegi, G.; Abraham, N.L.; Bowman, K.W.; Pyle, J.A.; Shindell, D.T. Global multi-year O$_3$-CO correlation patterns from models and TESsatellite observations. *Atmos. Chem. Phys.* **2011**, *11*, 5819–5838.

36. Wang, Y.; Hao, J.; Mcelroy, M.B.; Munger, J.W.; Ma, H.; Nielsen, C.P.; Zhang, Y. Year round measurements of O$_3$ and CO at a rural site near Beijing: Variations in their correlations. *Tellus B* **2010**, *62*, 228–241. [CrossRef]

37. Reddy, M.S.; Venkataraman, C. Inventory of aerosol and sulphur dioxide emissions from India. Part II—Biomass combustion. *Atmos. Environ.* **2002**, *36*, 699–712. [CrossRef]

38. Xu, W.Y.; Zhao, C.S.; Ran, L.; Lin, W.L.; Yan, P.; Xu, X.P. SO$_2$ noontime-peak phenomenon in the North China Plain. *Atmos. Chem. Phys.* **2014**, *14*, 7757–7768. [CrossRef]

39. Walker, H.L.; Heal, M.R.; Braban, C.F.; Ritchie, S.; Conolly, C.; Sanocka, A.; Dragosits, A.; Twig, G.M.M. Changing supersites: Assessing the impact of the southern UK EMEP supersite relocation on measured atmospheric composition. *Environ. Res. Commun.* **2019**, *1*, 041001. [CrossRef]

40. Zou, Y.; Deng, X.J.; Zhu, D.; Gong, D.C.; Wang, H.; Li, F.; Tan, H.B.; Deng, T.; Mai, B.R.; Liu, X.T.; Wang, B.G. Characteristics of 1 year of observational data of VOCs, NO$_x$ and O$_3$ at a suburban site in Guangzhou, China. *Atmos. Chem. Phys.* **2015**, *15*, 6625–6636. [CrossRef]

MDPI
St. Alban-Anlage 66
4052 Basel
Switzerland
Tel. +41 61 683 77 34
Fax +41 61 302 89 18
www.mdpi.com

Atmosphere Editorial Office
E-mail: atmosphere@mdpi.com
www.mdpi.com/journal/atmosphere